Synthetic Biology and Biotechnology: Safety, Security and Risk Assessment

Synthetic Biology and Biotechnology: Safety, Security and Risk Assessment

Quentin Webster

Larsen & Keller
www.larsen-keller.com

Synthetic Biology and Biotechnology: Safety, Security and Risk Assessment
Quentin Webster
ISBN: 979-8-88836-057-6 (Hardback)

Larsen & Keller
Published by Larsen and Keller Education,
5 Penn Plaza,
19th Floor,
New York, NY 10001, USA

Cataloging-in-Publication Data

Synthetic biology and biotechnology : safety, security and
risk assessment / Quentin Webster.
p. cm.
Includes bibliographical references and index.
ISBN 979-8-88836-057-6
1. Synthetic biology. 2. Biotechnology. 3. Biochemical engineering.
4. Biosynthese. I. Webster, Quentin.
TA164 .S96 2023
660.6--dc23

For more information regarding Larsen and Keller Education and its products, please visit the publisher's website www.larsen-keller.com

Table of Contents

Preface

Synthetic biology is a branch of science that seeks to create new biological parts, devices and systems, or to redesign systems that already exist in nature. It is a multidisciplinary area of research that encompasses a wide range of methodologies from several disciplines including biotechnology, molecular biology, chemical and biological engineering, etc. There are two different approaches in synthetic biology, namely, top-down approach and bottom-up approach. Synthetic biology researches are being used extensively for developing sustainable solutions for problems related to the fields of medicine, manufacturing and agriculture. In bioprocess technology, synthetic biology is used to achieve the production of high-value compounds important for vital biological activities. This book outlines the concepts and applications of synthetic biology and biotechnology in detail. Some of the diverse topics covered herein address the varied aspects of safety, security and risk assessment with respect to this field. The book is appropriate for students seeking detailed information in this area as well as for experts.

This book unites the global concepts and researches in an organized manner for a comprehensive understanding of the subject. It is a ripe text for all researchers, students, scientists or anyone else who is interested in acquiring a better knowledge of this dynamic field.

I extend my sincere thanks to the contributors for such eloquent research chapters. Finally, I thank my family for being a source of support and help.

Quentin Webster

1

Synthetic Biology and Biosecurity Concerns

Benjamin D. Trump, Marie-Valentine Florin, Edward Perkins and Igor Linkov

1.1 An Introduction to the Dual-Use Nature of Emerging Biotechnology

Synthetic biology uses engineering-based modeling and building techniques to modify existing organisms and microbes or to construct them from scratch. The rate of development and research related to synthetic biology for both industry and academia has increased over the past two decades (Ahteensuu 2017), with applications in medicine (new vaccines, delivery of therapeutics, and treatments), energy (biofuels), environmental remediation, food production, and general industry (detergents, adhesives, perfumes) (Evans and Selgelid 2015; Gronvall 2015).

While synthetic biology heralds advances in these fields, its techniques could also be adapted for malicious purposes and used by terrorist organizations, rogue actors, or hostile nations to create dangerous pathogens, invasive organisms, or other disruptive biological agents (Yeh et al. 2012). Such potential makes synthetic biology a dual-use research area of concern (DURC) since the same techniques can be used to benefit or harm people, animals, environments, or nations (Getz and

B. D. Trump (✉) · I. Linkov
US Army Corps of Engineers, Engineer Research and Development Center,
Concord, MA, USA
e-mail: Benjamin.D.Trump@usace.army.mil; igor.linkov@usace.army.mil

M.-V. Florin
International Risk Governance Center, École Polytechnique Fédérale de Lausanne,
Lausanne, Switzerland
e-mail: Marie-Valentine.Florin@epfl.ch

E. Perkins
US Army Corp of Engineers, Engineer Research and Development Center,
Vicksburg, MS, USA
e-mail: edward.j.perkins@usace.army.mil

Dellaire 2018). Indeed, there have been 35 confirmed cases of biological weapons deployment between 1970 and 2014 (Franconi et al. 2018). Thus far, the use of advanced biotechnologies for weapons production has mostly been pursued by state-actors. However, synthetic biological weapons are expected to become a larger concern as the field advances and they become increasingly adopted by malicious sub-state or non-state actors (Gronvall 2018).

Synthetic biology raises the possibility that pathogenic bioweapons could be designed, developed, and deployed in new ways that diverge from the disease-causing characteristics of naturally occurring pathogens (NAS 2018). Traditionally, only known pathogens found naturally in the environment, such as *B. anthracis* and *Y. pestis*, were developed as biological weapons because of the inherent infectious characteristics that readily enabled such organisms to serve as weapons. However, as synthetic biology continues to expand its capabilities to create and modify biological weapons, there is an increasing need for biosafety and biosecurity assurances for humans, animals, plants, and the environment. To cope with threats arising from synthetic biology's dual-use nature, biosecurity is needed to prevent, detect, and determine the source of biological attacks.

Biological weapons created from synthetic biology represent a new and unique threat space. Potential threats from synthetic biology include increased pathogen transmissibility between and within species in addition to resistance to established treatments. Synthetic biology can be used to engineer normally benign microbes that produce toxic biological compounds or re-build extinct or hard-to-obtain pathogens from scratch (NAS 2018). The main biotechnologies of concern in the near future (over the course of the next decade or so) are projected to be oligonucleotide synthesis, DNA assembly (assembling multiple smaller fragments of oligonucleotides into the desired larger sequence), and genetic modification (editing, deleting, and inserting desired sequences into targeted sites of a genome). Harm may also arise through the purposeful use of these techniques to disrupt human and environmental systems. Malevolent use, but also negligent use (misuse) of synthetic biology techniques, require two circumstances: (a) the spread of information, techniques, or knowhow to utilize synthetic biology's enabling technologies for irresponsible or nefarious purposes ("information hazard"), and (b) the ability to use such knowledge and tools to generate and disseminate harmful engineered organisms to vulnerable recipients.

Creating effective biosecurity procedures and policies to protect humans, agriculture, technology, and the environment from such nefarious usages (or accidental or negligent damage from misuse) will require understanding the current state of synthetic biology. This includes knowing the platforms and technologies available for manipulation or construction (e.g., viruses, microbes, multi-cellular organisms, or cell-free systems) and planning for the future as the field overcomes identified bottlenecks or roadblocks. Biosecurity will require developing screening mechanisms for synthetic pathogens and biological attacks, as well as methods to assess how a synthetic pathogen could be dispersed during an attack, identify what the potential targets of the attack are, and identify who developed the bioweapon.

1.2 History

State-sponsored biological warfare has been employed since ancient times (Mayor 2003). By 1925, germ theory and advances in bacteriology had led to enough advancement in the field that regulation was deemed necessary, leading to the Geneva Protocol, which banned the use of chemical and biological weapons in war. However, this did not prevent a number of countries from embarking on biological weapons programs. A notorious usage occurred during World War II, when Japanese units deployed bombs filled with plague-ridden fleas in China as a part of their biological weapons program. Other nations also invested in biological weapons programs, including the UK and the US, although they did not end up deploying the weapons they developed. These state-based biological weapons capabilities tended to be justified as "first strike" options or as methods to weaken enemy ground forces that could not be combatted through conventional ordnance. In all cases, states developed the scientific and material ability to build, package, and deploy biological weapons.

A major regulatory development came in 1972, with the passing of the Convention on the Prohibition of the Development, Production and Stockpiling of Bacteriological (Biological) and Toxin Weapons and on Their Destruction (usually referred to as the Biological Weapons Convention; here "BWC"), an update to the Geneva Protocol. The BWC banned the "development, production and stockpiling of microbes or their poisonous products except in amounts necessary for protective and peaceful research" ("Convention," 1972). The BWC went further than the Geneva Protocol in banning not just the use of biological weapons in war, but also their development and possession. Although 183 countries are parties to the BWC, some countries with a known interest in biological weapons, such as Israel, have not signed or ratified the BWC, and some countries that have ratified the BWC, such as Russia and China, have completely or partially ignored its statutes. Russia had an active biological weapons program comprising tens of facilities and thousands of researchers until at least 1992, twenty years after it signed the BWC (Zilinskas 2012). China is believed to have aided Iran's biological weapons program through the sale of sensitive dual-use equipment and vaccines since at least 1997 (Byman et al. 1999). Under the BWC, member nations agree to not aid the biological weapons programs of foreign nations, thereby making China's past trade actions noncompliant with the BWC. However, the BWC lacks mechanisms to enforce compliance and to investigate and respond to events of noncompliance.

Nevertheless, for almost fifty years the BWC has been the gold standard for watching for and preventing state-sponsored biological weapons. Advancements in synthetic biology have decreased the amount of time, money, and skill needed for an individual or organization to participate in the field, leading to the development of pathogens by smaller sub-state and non-state actors who are not restricted by the BWC (Evans and Selgelid 2015), as well as potentially by individuals with relatively simple scientific knowledge. For example, al-Qaeda and Aum Shinrikyo have both attempted to develop bioweapons, although with limited success (Tucker

2007). In the 2001 "Amerithrax" bioterror attack, an individual actor sent *Bacillus anthracis* spores through the U.S. Postal Service, resulting in five deaths, the prophylaxis of 30,000 more individuals, and hundreds of millions of dollars in decontamination expenses (Pita and Gunaratna 2010). The destructive potential of synthetic biology is only increasing with scientific advances in the field. In the future, threats could come from disgruntled employees of state-of-the-art scientific laboratories or vengeful academics, but also simply from a sorcerer's apprentice, insufficiently prudent and educated to prevent accidents. Therefore, as the newest developments in genetic engineering grow increasingly accessible to private citizens, the question arises as to what rate-limiting steps (e.g. technical knowledge or inspiration) can prevent non-state small group or individual actors from undertaking development of increasingly advanced biological weapons.

Indeed, now the average person interested in synthetic biology can set up a lab in their home, using common household items and equipment or kits ordered over the internet, in a movement now known as "DIY Bio" ("do-it-yourself biology") or "garage biology" (Ledford 2010). Additionally, while chemicals have a set structure and formula that makes them easier to identify and govern, advancements in synthetic biology allow for the creation of novel types of biological weapons that are not explicitly prohibited by the BWC but instead exist in a sort of "gray zone." This increased accessibility of synthetic biology and difficulty in categorizing synthetic biology products creates a threat space that is not fully captured within the current structure or implementation of the BWC.

1.3 Effective Governance and Policy for Biosecurity

Because emerging biotechnologies are dual-use, governance must weigh the risk of misuse with the potential for beneficial use in innovation and development. Unfortunately, biosecurity attempts are mired in uncertainty around both the actual capabilities of synthetic biology, as well as the motivations of actors given the increasing number of contexts in which synthetic biology is used. Modern governments are still relying on old rules to regulate a new technology, clearly an insufficient strategy for ensuring security in the coming decades.

Building an effective biosecurity strategy to encompass twenty-first century biotechnologies requires understanding the novelties that sciences like synthetic biology create in the biosecurity threat space, as well as the structural vulnerabilities these sciences can exploit and the likely causes of inadequate biosecurity practices. Synthetic biology's novel biosecurity concerns arise from its broad scope, wider availability, complexity, and uncertainty over current and future capabilities. For example, critical developments such as gene editing via CRISPR gene editing vastly improve upon previous genetic engineering processes and may yield a revolution in human and environmental health research, but may also cause substantial and irreversible harms. One application of gene editing is the gene drive, which can rapidly propagate a certain set of genes or alleles through a population, circumventing

Mendelian inheritance laws and increasing the chance that this set of genes is passed on. While gene drives are an exciting new technology, their ability to rapidly alter the genetic makeup of a population is cause for concern. Other potential negative consequences of gene editing may include the unconstrained diffusion of gene-edited material throughout the environment, the disruption of ecologies with genetically-modified organisms (in particular engineered gene drive systems), and off-target impacts from genome editing. These techniques could also be used maliciously, with an actor purposely targeting humans and/or the environment.

The publication and dissemination of a methodology for synthesizing horsepox in a laboratory setting was a recent application of gene editing (Noyce et al. 2018). Some critics say this information could support a nefarious actor to reconstitute and develop smallpox, or to synthetize other viruses. Additionally, the widely publicized recreation of the 1918 Spanish Influenza (Tumpey et al. 2005), which killed some 50 million people worldwide at the close of the First World War, could facilitate the synthesis process for actors wishing to cause harm. Even nonpathogenic approaches have been described as dual-use research, ranging from the disruption of local ecologies via gene drives to the manipulation or destruction of inorganic materials.

These and dozens of other cases demonstrate the increasing ease with which an actor can acquire information and apply existing tools to deploy advanced genetic engineering applications with limited to no oversight. In 1975, the U.S. National Institute of Health ("NIH") established compliance measures for genome engineering that were enforced through funding restrictions; however, many synthetic biology innovators can now operate without NIH funding, approval, or even awareness, and NIH does not oversee research in other countries. Today, the financial costs, time limitations, and skill requirements needed to wield synthetic biology tools have scaled down such that some of these tools have become accessible even to elementary school students. Furthermore, the requisite baseline knowledge diminishes over time as synthetic biology processes become more streamlined. While such broad access to sophisticated genetic engineering knowledge and equipment can accelerate scientific breakthroughs, it also places the responsibility for biosecurity on a near infinite number of unsupervised actors across the globe.

In 2018, the BWC secretariat noted that increased access to technologies such as gene editing, gene drives, and gene synthesis is available to actors with limited or no oversight from established industry or governmental organizations, raising concerns about potential violations of the BWC. It is helpful to forecast and understand looming threats and potential mitigation strategies at various scales, but international treaties are not structured to oversee bottom-up efforts related to the localization and globalization of synthetic biology below the national scale. One part of the solution may be the broadening of engagement from established oversight agencies like NIH.

An additional option is the Responsible Research and Innovation ("RRI") approach, utilized by the UK and the EU, which appraises the potential effects of new research on society and the environment in order to improve the alignment of processes and expected outcomes with societal values and needs. RRI approaches include experts from a range of different fields whose role it is to assess scientific development with the goal of mitigating risk, making research advances accessible through fair and sustainable means, and upholding key morals and values. Programs that adopt the RRI approach are not meant to prevent research or the publication of results, but rather to minimize downstream harms that could make developers, companies, and/or governments liable for costly insurance and cleanup efforts. RRI becomes an important criteria in access to public funding but is not a regulatory requirement.

Biosecurity could also indicate to the general public that certain synthetic biology products have been filtered to guarantee beneficial uses. The US Nuclear Regulatory Commission (nrc.gov), for instance, performs this role for research and test reactors; the synthetic biology field would benefit from a similar regulatory body for biosecurity.

Where top-down governance proves insufficient, other actors such as universities, non-profits, and companies will need to engage their own gatekeeping and watchdog capabilities to protect against nefarious actors. Top-down governance may support such initiatives, which will require harmonization and communication up to the international level. These initiatives will need to be incentivized. Currently, though, biosecurity is viewed as an obligation, such that individuals, organizations, and companies must use their own funds to observe unstated and often confusing or contradictory needs for overall security. This balance of costs and benefits is insecure, and as such, institutions tend to want to minimize expenditures associated with oversight (Gillum et al. 2018). The best argument to support investing in biosecurity is that the advancement of synthetic biology ultimately requires public approval, whereas currently the public remains quite skeptical (Pauwels 2013; Oliver 2018). The public could grow more opposed to synthetic biology were the public inadvertently exposed to some harm as a consequence of insufficient or inadequate oversight. Biosecurity necessitates a strategy which incentivizes managers and corporations to stay up-to-date with the latest risks and concerns.

Some corporations are aware of the risks to their bottom line should the public be exposed to harm arising from a synthetic biology product. The majority of DNA synthesis companies, in fact, have joined the International Gene Synthesis Consortium ("IGSC"), whereby they utilize company resources to monitor customers and their requests for potential security problems. They are aware that implementing such biosecurity measures is in their best interests, even though there is no legal regime requiring them to do so. Likewise, in January 2020, the World Economic Forum and the Nuclear Threat Initiative published a report recommending that a technical consortium be set up in order to create a common DNA sequence screening mechanism. This screening mechanism would be based on work done by the IGSC.

Viewing longstanding biosecurity policy practices through the lens of risk analysis results in the conclusion that there are significant gaps in biosecurity effectiveness for synthetic biology. These inefficient and inadequate policies include (a) viewing security as a cost or undesirable expense to be minimized, (b) the siloing of scholarship and practice across disciplinary domains and among government, industry, academia, and civil society, and (c) the narrow framing of security problems ignores new actors and technological developments taking place in a variety of different countries and in adjacent technology fields. Each of these issues could be resolved through policy solutions that both encourage technological development and mitigate security threats while enabling public engagement in synthetic biology and investment in its products as they enter the marketplace. Policies for synthetic biology must be scalable, transferable, and adaptable in order to take into account its emerging technical and social challenges.

The increasingly globalized, distributed, and dispersed nature of synthetic biology products and research worsens challenges arising from differing practices of biosecurity governance globally. Advanced biological research is no longer overwhelmingly dominated by Europe and the US, and this may introduce different approaches to, or priorities for, biosecurity. Russia's Federal Research Programme for Genetic Technologies Development for 2019–2027, for instance, intends to "implement a comprehensive solution to the task of the accelerated development of genetic technologies, including genetic editing; to establish scientific and technological groundwork for medicine, agriculture and industry; to improve the system of preventing biological emergencies and monitoring in this area" (Ministry of Science and Higher Education of the Russian Federation 2019). Similarly, Saudi Arabia is funding research related to the development of microbial cell factories to produce fuels and chemicals, while the Singaporean government is investing considerable resources into the funding of life and environmental sciences researchat Nanyang Technological University, the National University of Singapore, and the Agency for Science, Technology and Research (A*STAR). The Chinese Academy of Sciences is establishing an Institute of Synthetic Biology, which is tasked with the dual responsibilities of fostering roadmaps for the future development of Chinese synthetic biology while also establishing safety and security norms for researchers at Chinese institutions. There are no top-down efforts beyond existing mechanisms like the BWC or the CWC (The Convention on the Prohibition of the Development, Production, Stockpiling and Use of Chemical Weapons and on their Destruction) that standardize global governance and usage of synthetic biology, and bottom-up efforts are not coordinated in their reach or messaging.

Relative newcomers to the development of synthetic biology may possess differing tolerances and constructions of risk compared to more established technology developers. The implications of the entry of such newcomers to the field, though vast, can be grouped into two general areas. One includes diverging safety and security practices at various points of an international supply chain that forms the backbone of an increasingly globalized economy. Another includes the potential for small-scale experiments or national biosecurity policies to escape the given actor's control and spill across political boundaries. While one country may find the environmental risk of a particular synthetic biology application acceptable, its spread across borders into another country may disrupt those local ecologies (i.e., crashing or hardening a particular species through genetic engineering) or expose vulnerable human populations to irreversible consequences without options for amelioration. The nature of certain synthetic biology applications (i.e. gene drives) makes it impossible for risk-averse countries to wholly quarantine themselves from exposure to harms resulting from another country's decisions. This is also an issue of equity given that risk-tolerant countries will reap the rewards of risks when beneficial technologies emerge, but risk-averse countries may bear their neighbors' risks without any means to capture potential rewards.

An environment of competing and incongruent risk architectures causes individual states, organizations, or industries to arrive at differing definitions of security threats or acceptable levels of loss in pursuit of a technology's intended gains. For a technology as uncertain as synthetic biology, this policy divergence may set governments, companies, and other research organizations down vastly differing policy paths, and impede consensus in assessing the minutiae of technical risk concerns or assessment protocols, or ensuring security for anyone.

1.4 Conclusion

Synthetic biology is a transformative technology with the possibility to change the world to the same extent as – if not more than – the digital revolution. As is the case with previous scientific breakthroughs, the potential for its dual-use and misuse represents a global problem, and necessitates that the highest levels of policy makers pay it close attention. Although targeted countermeasures can go some way toward providing protection, preventative actions are likely to be more effective given the heightened uncertainty of the field's future (Trump et al. 2020a). It is necessary that biosecurity policies and practices be updated to take into account both the unprecedented challenges associated with synthetic biology and the globalized, diffuse, and varied nature of its threat space.

Effective global biosecurity will not happen quickly, nor will it be enthusiastically adopted by all national governments and non-governmental organizations. Incentives to misuse synthetic biology with harmful consequences remain high for certain negligent actors, as are the incentives to dual-use by nefarious actors, and the coming years may see such events affecting human or environmental health. Successful biosecurity implementation must be adaptable to quickly incorporate uncertainty as well as new capabilities. Urgent steps are required to place such notions into practice before a major threat incident, which in addition to creating substantial damages could usher in policy changes that might limit or ban platforms of synthetic biology research entirely. Besides forgoing the benefits of the technology, such a ban could force development underground and further out of the reach of coordinated governance or risk assessment. By identifying the social, economic, institutional, and technological tripwires that influence a state's trajectory towards biosecurity research of concern, now is the time to take steps to apply biosecurity to maximize technological benefits while minimizing the dual-use potential of synthetic biology by improving the framing, prioritization, and governance of biosecurity risks.

Many individuals and organizations are already taking on the challenge of biosecurity. For instance, the International Genetically Engineered Machine (iGEM) synthetic biology competition requires that judges carry out rigorous reviews of each team's planned experiments and the materials they will require. Additionally, iGEM's Safety and Security Committee as well as a commercial partner review team plans for potential harms (McNamara et al. 2014; Millet et al. 2019).

Biosecurity is prioritized by a number of independent and governmental organizations, including the DIY biology movement, which has a code of ethics operative in North America and in Europe (DIYbio.org). This prioritization is also indicated by the intent of the member states of the BWC to develop a code of conduct (Meeting of the States Parties 2018), the African Union's development of biosecurity norms for transgenic insects and genetically-modified crops (Glover et al. 2018), and the MIT Bio Summit 2.0 (www.biosummit.org) statement of shared purpose. Additionally, there is increasing demand for an update to global biosecurity practices and norms along the lines of the Cartagena Protocol on Biosafety. Such an update should improve cooperation, transparency, and joint security in the practice of synthetic biology.

Still, the problem of how to incentivize private actors to invest in biosecurity remains. The answer will require the participation not only of bench scientists, but also of various overseers, gatekeepers, and watchdog groups involved in biotechnology research and development (for instance, the World Organisation for Animal Health's Guideline for Responsible Conduct in Veterinary Research). One example of a potential approach is to train journal editors to recognize potential information hazards within article submissions. Additionally, funders responsible for reviewing grants could require that applicants include a review of potential information and security hazards which might occur over the course of the proposed work. In these and other examples, a fusion of top-down and bottom-up approaches is necessary in order to identify security threats and to raise awareness of biosecurity issues; meanwhile, bottom-up organizations can develop on-the-ground passive surveillance programs to monitor potential dual-use security threats.

An example of this collaboration is between the US Federal Bureau of Investigation (FBI) and iGEM. The FBI has funded iGEM and collaborated with the competition's organizers in order to increase awareness of risk and to build an understanding of possible or developing threats. While no biosecurity policy or practice can completely eliminate all threats (especially not without unilaterally preventing research and innovation that could greatly benefit society), a layered approach in which government, private organizations, and individual citizens collaborate will result in a more unified effort for biosecurity which could reduce gaps in oversight that might be exploited by actors looking to develop biological weapons.

Ultimately, adapting national and global biosecurity principles and practices may yield a number of downstream benefits that will help emerging biotechnologies become safer, more responsibly available, insurable, and trusted by key stakeholders and the broader public. Actions to achieve such goals, such as promoting multistakeholder discourse for top-down and bottom-up governance, increased global coordination, and crafting and implementing mechanisms that foster effective, sustainable, and adaptive biosecurity, are likely needed to bridge the gap between earlier generation biotechnology policy and the cutting-edge scientific and technological capabilities of the next decade (Trump et al. 2020a, b). This book includes a number of international opinions and scholarship on how such goals might be achieved, building from insights related to governance, risk assessment, ELSEI (ethical, legal, social and environmental issues), forecasting and horizon scanning, and the state of the art in physical and life sciences. While solutions for biology will be difficult to achieve due to a number of competing incentives, political challenges, and institutional requirements, we believe that the ideals expressed herein help signal how improved biosecurity might be achieved over time.

To explore the various topics pertaining to emerging biosecurity challenges, the chapters that follow build from a NATO Science for Peace and Security Programme (SPS) Advanced Research Workshop (ARW) hosted by the École Polytechnique Fédérale de Lausanne (EPFL) in Lausanne, Switzerland in July 2019. Specifically, participants from over two dozen countries gathered to explore various physical and life sciences as well as social sciences concerns related to emerging biotechnology and associated security needs, challenges, and opportunities. This book includes written perspectives from the various working groups, including (a) top-down governance at the national and international levels, (b) bottom-up governance using grassroots and self-governing interests, (c) the uncertain and difficult challenges posed by information hazards, (d) security concerns from technical, life sciences, and computer-assisted design research, and (e) the approaches and challenges of foresight in the forecasting of future threats, opportunities, and governance needs within the broader biosecurity space (Linkov et al. 2018). Additional chapters emphasize more explicit topics within each of these core ideas, and elucidate different perspectives on how to better understand the limitations of modern biosecurity, as well as how it may be improved from a risk assessment, governance, and multidisciplinary sciences perspective.

References

Ahteensuu M (2017) Synthetic biology, genome editing, and the risk of bioterrorism. Sci Eng Ethics 23:1541–1561. https://doi.org/10.1007/s11948-016-9868-9

Byman D, Cliff R, Saunders P (1999) US policy options toward an emerging China. Pac Rev 12(3):421–451

Convention on the Prohibition of the Development, Production and Stockpiling of Bacteriological (Biological) and Toxin Weapons and on Their Destruction. Biological and Toxin Weapons Convention Website. 10 April 1972. http://www.opbw.org/

Evans NG, Selgelid MJ (2015) Biosecurity and open-source biology: the promise and peril of distributed synthetic biological technologies. Sci Eng Ethics 21:1065–1083. https://doi.org/10.1007/s11948-014-9591-3

Franconi R, Illiano E, Paolini F, Massa S, Venuti A, Demurtas OC (2018) Rapid and low-cost tools derived from plants to face emerging/re-emerging infectious diseases and bioterrorism agents. In: Defence against bioterrorism. Springer, Dordrecht, pp 123–139

Getz LJ, Dellaire G (2018) Angels and devils: dilemmas in dual-use biotechnology. Trends Biotechnol 36(12):1202–1205

Gillum D, Carrera LAO, Mendoza IA, Bates P, Bowens D, Jetson Z, Maldonado J, Mancini C, Miraldi M, Moritz R (2018) The 2017 Arizona biosecurity workshop: an open dialogue about biosecurity. Appl Biosaf 23:233–241. https://doi.org/10.1177/1535676018781854

Glover B, Akinbo O, Savadogo M, Timpo S, Lemgo G, Sinebo W, Akile S, Obukosia S, Ouedraogo J, Ndomondo-Sigonda M (2018) Strengthening regulatory capacity for gene drives in Africa: leveraging NEPAD's experience in establishing regulatory Systems for Medicines and GM crops in Africa. BMC Proc 12(Suppl 8):11. https://doi.org/10.1186/s12919-018-0108-y
Gronvall G (2015) Biosecurity: the opportunities and threats of industrialization and personalization. Bull At Sci 71(6):39–44
Gronvall GK (2018) Safety, security, and serving the public interest in synthetic biology. J Ind Microbiol Biotechnol 45:463–466. https://doi.org/10.1007/s10295-018-2026-4
Ledford H (2010) Life hackers. Nature 467(7316):650
Linkov I, Trump BD, Anklam E, Berube D, Boisseasu P, Cummings C, Ferson S, Florin M-V, Goldstein B, Hristozov D (2018) Comparative, collaborative, and integrative risk governance for emerging technologies. Environ Syst Decis 38:170–176. https://doi.org/10.1007/s10669-018-9686-5
Mayor A (2003) Greek fire, poison arrows & scorpion bombs: biological and chemical warfare in the ancient world. Overlook Duckworth, Woodstock. ISBN:978-1-58567-348-3
McNamara J, Lightfoot SB-Y, Drinkwater K, Appleton E, Oye K (2014) Designing safety policies to meet evolving needs: iGEM as a testbed for proactive and adaptive risk management. ACS Publications, Washington, DC
Meeting of the States Parties (2018) Meeting of experts on review of developments in the field of science and technology related to the convention: reflections and proposals for possible outcomes submitted to the meeting of the states parties. Switzerland, Geneva
Millett P, Binz T, Evans SW, Kuiken T, Oye K, Palmer MJ et al (2019) Developing a comprehensive, adaptive, and international biosafety and biosecurity program for advanced biotechnology: the IGEM experience. Appl Biosaf 24(2):64–71
Ministry of Science and Higher Education of the Russian Federation. Approval of the Federal Research Programme for Genetic Technologies Development for 2019–2027. The Russian Government: Government Decisions Website. 22 April 2019. http://government.ru/en/docs/36457/
National Academies of Sciences, Engineering, and Medicine (2018) Biodefense in the age of synthetic biology. National Academies Press, Washington, DC. https://doi.org/10.17226/24890
Noyce RS, Lederman S, Evans DH (2018) Construction of an infectious horsepox virus vaccine from chemically synthesized DNA fragments. PLoS One 13(1):e0188453. https://doi.org/10.1371/journal.pone.0188453
Oliver A (2018) Behavioral economics and the public acceptance of synthetic biology. Hastings Cent Rep 48:S50–S55
Pauwels E (2013) Public understanding of synthetic biology. Bioscience 63:79–89. https://doi.org/10.1525/bio.2013.63.2.4
Pita R, Gunaratna R (2010) Anthrax as a biological weapon: from World War I to the Amerithrax investigation. Int J Intell Counterintell 23(1):61–103
Trump BD, Cummings CL, Kuzma J, Linkov I (2020a) Synthetic biology 2020: frontiers in risk analysis and governance. Springer, Cham
Trump BD, Galaitsi SE, Appleton E, Bleijs DA, Florin MV, Gollihar JD et al (2020b) Building biosecurity for synthetic biology. Mol Syst Biol 16(7):e9723
Tucker JB (2007) War of nerves: chemical warfare from World War I to Al-Qaeda. Anchor, New York
Tumpey TM, Basler CF, Aguilar PV, Zeng H, Solorzano A, Swayne DE, Cox NJ, Katz JM, Taubenberger JK, Palese P, Garcia-Sastre A (2005) Characterization of the reconstructed 1918 Spanish influenza pandemic virus. Science 310(5745):77–80. https://doi.org/10.1126/science.1119392
Yeh JY, Seo HJ, Park JY, Cho YS, Cho IS, Lee JH et al (2012) Livestock agroterrorism: the deliberate introduction of a highly infectious animal pathogen. Foodborne Pathog Dis 9(10):869–877
Zilinskas RA (2012) Take Russia to 'task' on bioweapons transparency. Nat Med 18(6):850–850

2

Literature Review of Biosecurity Threats and Responses

Christopher L. Cummings, Kaitlin M. Volk, Anna A. Ulanova, Do Thuy Uyen Ha Lam and Pei Rou Ng

2.1 Introduction

The field of biotechnology has been rigorously researched and applied to many facets of everyday life. Biotechnology is defined as the process of modifying an organism or a biological system for an intended purpose. Biotechnology applications range from agricultural crop selection to pharmaceutical and genetic processes (Bauer and Gaskell 2002). The definition, however, is evolving with recent scientific advancements. Until World War II, biotechnology was primarily siloed in agricultural biology and chemical engineering. The results of this era included disease-resistant crops, pesticides, and other pest-controlling tools (Verma et al. 2011). After WWII, biotechnology began to shift domains when advanced research on human genetics and DNA started. In 1984, the Human Genome Project (HGP) was formerly proposed, which initiated the pursuit to decode the human genome by the private and academic sectors. The legacy of the project gave rise to ancillary advancements in data sharing and open-source software, and solidified the prominence of "big science;" solidifying capital-intensive large-scale private-public

C. L. Cummings (✉)
North Carolina State University, Raleigh, NC, USA

Iowa State University, Ames, IA, USA

K. M. Volk · A. A. Ulanova
US Army Corps of Engineers, Environmental Laboratory, Engineer Research and Development Center, Concord, MA, USA

D. T. U. H. Lam
Genome Institute of Singapore (GIS), Agency for Science, Technology and Research (A*STAR), Singapore, Singapore

P. R. Ng
Wee Kim Wee School of Communication and Information, Nanyang Technological University, Singapore, Singapore

research initiatives that were once primarily under the purview of government-funded programs (Hood and Rowen 2013). After the HGP, the biotechnology industry boomed as a result of dramatic cost reduction to DNA sequencing processes. In 2019 the industry was globally estimated to be worth $449.06 billion and is projected to increase in value (Polaris 2020).

While biotechnology is lauded for its anticipated positive impacts on society, new public health challenges are also likely given the scientific and technological advances made in areas like bioengineering and gene editing (Trump et al. 2020a). Misuse of powerful biotechnologies is of significant concern, be it purposeful or accidental. For instance, the 1979 Sverdlovsk anthrax leak occurred when soviet scientists accidentally released genetically modified microorganisms from their biological weapons facility. The incident resulted in over 100 casualties in nearby populations (Sahl et al. 2016). This case not only demonstrates tragic consequences of biotechnological misuse but also highlights purposeful negligence and gross impertinence regarding international agreements, in this case the 1972 Biological Weapons Convention (BWC). The BWC was an agreement signed by 183 countries that banned biological weapons by countries' self-regulated accord to prohibit the development, production and stockpiling of biological agents or related equipment that could realize a biological attack (UNODA 2017). Since the 1970s, threats posed by biotechnological tools have become arguably more widespread as production costs have decreased while access to processing tools have increased. Technologies such as CRISPR and RT-PCR are available in many academic and research laboratories, increasing the possibility of independent actors misusing the technology for nefarious purposes. Increased access and ease of use also correlates to a greater diversity of individuals using biotechnology tools for distinct purposes—some of which are deemed unethical or antithetical to global standards for biotechnology research and application. An infamous example is the 2018 experiment that resulted in the birth of two twin girls where a research group applied CRISPR technology to immunize the embryos against HIV. Although this isolated incident did not pose direct national security threats, the experiment does open up a "Pandora's box" of possible unethical misuses (Raposo 2019). Following this pernicious history, there is a demonstrated need for the development and synthetization of a coordinated biotechnology framework that can better prioritize and anticipate biotechnological risks while seeking to maximize the potential benefits of applications.

Biosecurity frameworks' essential function is to create a protocol that minimizes the collateral damage of pathogens and pests. The BWC is a keystone of international biosecurity policy that arose out of the need to protect nations from the threat of an engineered biological attack. The US Department of Agriculture defines biosecurity as the methods and procedures intended to "prevent the introduction, delivery, and spread of disease pathogens that can harm or adversely affect livestock, crops, environments and people," (USDA APHIS 2020). Similar to biotechnology, the principles behind biosecurity are based in agriculture and prioritize the need to protect monocultures of crops, livestock, and poultry, whose lack of genetic diversity makes them especially vulnerable to disease. As the biotechnology field progressed and new capabilities in gene sequencing, synthesis, and modification were refined, democratized, and globalized over the past decade, advanced biotechnology practices and products required greater prioritization of biosecurity practices and considerations.

Biosecurity threats include biological weapons and accidental releases as demonstrated in the Sverdlovsk anthrax event, but they have also become more diversified and complicated as researchers develop and utilize advanced biotechnology techniques for the betterment of society across other sectors. Gene drives for mosquito population control, engineered algae for biofuel creation, and recreation of extinct pathogens for novel vaccine development have unique and potentially unknown associated risks. The envisioned coordinated biosecurity framework would allow for beneficial innovation to proliferate while simultaneously reducing anticipated and unanticipated risk of harm to humans, animals, agricultural, and the environment (Trump et al. 2020b; Wells et al. 2020).

Many experts in the fields of public policy, public health, biotechnology, and more have discussed the threats that biotechnology may pose and the appropriate biosecurity responses from their diverse perspectives. To date, there has been no synthesis of published and gray literature regarding biosecurity. This chapter fills this gap in order to advance understanding of this quickly growing field. This chapter aims to define the typology of issues related to modern biosecurity threats and responses by coalescing disparate perspectives on biosecurity into a single descriptive location. In sum, we analyzed over one hundred peer-reviewed documents from 26 countries in order to identify reported threats and responses across global sources. The most prevalent threats identified in our analysis include dual use research of concern, biological weapons, and the ecological impact of advanced biotechnology products, while the most prevalent responses include regulation and legal oversight of the biotechnology field, risk assessment and management, frequent and open communication between researchers, government, industry, and the general public, and a strong adherence to ethics in the scientific community and subsequent self-governance. These threats and responses, in addition to less frequently mentioned ones, are discussed in this chapter.

2.2 Methods

We began our analysis with a systematic review of articles, where only published peer-reviewed articles (e.g., commentary, perspective, opinion, review articles) with available full-text were included. We started the systematic review process in the summer of 2019 and only articles published in the previous five years (2014 onwards) were considered for sampling as the area of interest is a recent emerging field (Fig. 2.1). The four databases we used to identify relevant articles were PubMed, Web of Science (WOS), Scopus, and ProQuest. We selected these databases as they provide a comprehensive collection of biomedical, life sciences, and social sciences articles. As we employed PubMed to execute the search, the Medical Subject Headings (MeSH) in PubMed were also utilized to acquire highly-specific

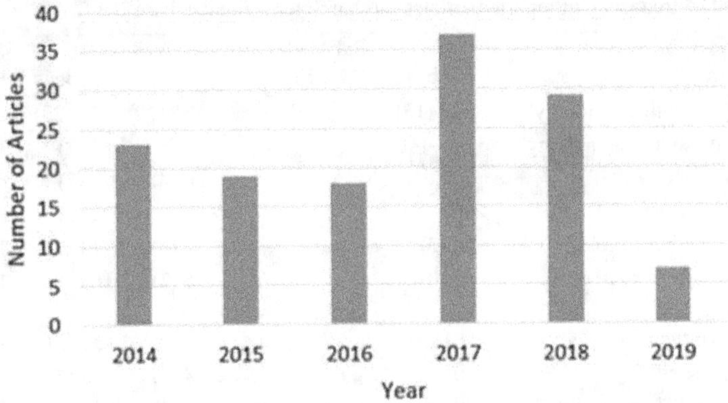

Fig. 2.1 Number of articles published between 2014 and summer of 2019 selected for our systematic review of advanced biotechnology biosecurity concerns

results based on specific medical search terms included below. The focus of our chapter is on the biosecurity threats and their proposed responses stemming from advanced biotechnology in the area of synthetic biology. Therefore, we selected search terms such that a wide range of relevant technologies were included: "gene drive", "virus", "micro", "gene edit", "CRISPR", "cell free", and "synthetic biology." To ensure thorough coverage of biosecurity issues, we employed different implicated terms. These terms included "biosecurity", "weapon", "defense", and "dual-use." We also applied different spellings (e.g. "gene edit*" and "bio-security") of the search terms to ensure that we obtained a comprehensive list of articles (Table 2.1).

First-level analysis included data "cleaning" to improve the relevance of the final sample of articles. We read each abstract provided by the databases to gauge the relevance of the article and to screen out any non-relevant results. We removed articles that addressed solely the technical aspect of advanced biotechnologies and articles that addressed other biosecurity concerns irrelevant to advanced biotechnologies. We also excluded papers with no or marginal discussion on biosecurity issues and/or measures, articles written in languages other than English, and articles in which the key words (e.g. "weapons"," defense") were used only metaphorically (Table 2.1). These criteria yielded 84 articles from PubMed (MeSH included), 119 articles from WOS, 145 articles from Scopus, and 184 articles from ProQuest. We then combined all of the articles gained from the four databases and removed any duplicates, bringing the total down to 166 articles. We were unable to obtain PDFs for 33 of these articles. Since this prevented us from analyzing the full content of these articles, we removed them from our review. The final number of articles included in our systematic review is 133.

In addition to the systematic review of peer-reviewed literature, we also sought to identify what different government agencies report about advanced biotechnologies and biosecurity and compare it with the common themes identified in the

Table 2.1 The frequency of reported articles by search term and database

Search term	PubMed	PubMed (MeSH)	WOS	Scopus	ProQuest
Biosecurity "synthetic biology"	27 **(13)**		35 **(20)**	40 **(23)**	26 **(5)**
Bio-security "synthetic biology"	0		0	2 **(2)**	1 **(0)**
Biosecurity "gene edit*"	0		6 **(4)**	7 **(7)**	67 **(7)**
Bio-security "gene edit*"	0		0	1 **(1)**	0
Weapon* "synthetic biology"	9 **(3)**		8 **(4)**	14 **(7)**	138 **(12)**
"Warfare agent*" "synthetic biology"	2 **(0)**		5 **(2)**	6 **(2)**	14 **(2)**
Defense "synthetic biology"	81 **(4)**		51 **(3)**	51 **(5)**	96 **(4)**
Weapon "gene edit*"	0		8 **(3)**	13 **(5)**	72 **(8)**
"Warfare agent*" "gene edit*"	0		3 **(3)**	3 **(3)**	7 **(1)**
Defense "gene edit*"	0		107 **(1)**	113 **(2)**	51 **(1)**
Biosecurity "dual-use"	36 **(27)**		32 **(20)**	57 **(39)**	83 **(31)**
Bio-security "dual-use"	0		0	4 **(1)**	4 **(3)**
Biosecurity "gene drive"	4 **(3)**		3 **(3)**	7 **(7)**	23 **(13)**
Bio-security "gene drive"	0		0	0	0
Biosecurity "micro*"	17 **(0)**		202 **(5)**	472 **(4)**	2625 **(26)**
Bio-security "micro*"	1 **(0)**		5 **(0)**	21 **(0)**	130 **(2)**
Biosecurity "cell-free"	2 **(0)**		1 **(0)**	1 **(0)**	44 **(1)**
Bio-security "cell-free"	0		0	0	2 **(0)**
Biosecurity "CRISPR"	22 **(8)**		10 **(9)**	11 **(8)**	103 **(27)**
Bio-security "CRISPR"	0		0	0	1 **(0)**
Biosecurity "virus"	1199 **(7)**		526 **(21)**	603 **(13)**	1707 **(39)**
Bio-security "virus"	11 **(0)**		9 **(0)**	29 **(2)**	87 **(2)**
Biosecurity "*virus"	1199 **(7)**		547 **(21)**	632 **(12)**	0
Bio-security "*virus"	11 **(0)**		9 **(0)**	30 **(2)**	0
Gene drive technology		13 **(12)**			

Non-bolded numbers give the frequency before initial screening of abstracts, while the bolded numbers in parentheses give the frequency after initial screening and abstracts
MeSH medical subject headings, *WOS* web of science

peer-reviewed literature. We focused on government documents published by the US and the EU as they dominated the academic conversation on biosecurity (Fig. 2.2). The National Academies of Sciences, Engineering, and Medicine's (2018) report *Biodefense in the Age of Synthetic Biology* was analyzed for the US and the European Commission's (2017) report *Action Plan to Enhance Preparedness against Chemical, Biological, Radiological and Nuclear Security Risks* was analyzed for the EU (Fig. 2.3).

Our next step was to analyze the content of each article. To do this, we read through each document and identified any biosecurity threats and responses (i.e.

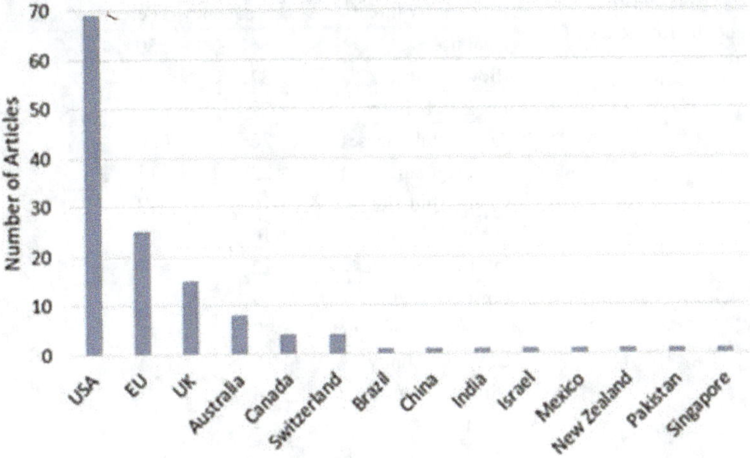

Fig. 2.2 The number of articles within the systematic literature review published from each country as determined by the affiliation of the corresponding author

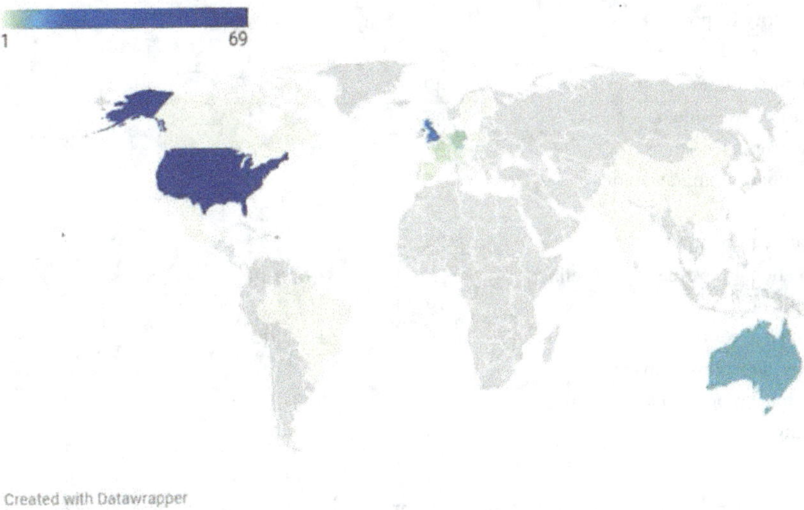

Fig. 2.3 Map displaying which countries published literature used in our systematic review and the relative frequency of publications from each country, as determined by the affiliation of the corresponding author

solutions to biosecurity threats) described in the document. We then grouped common threats and responses until we had a comprehensive and manageable list of biosecurity threats (Table 2.2) and expert-recommended responses (Table 2.3). This processes followed a grounded theory coding structure to ensure that we maintained a theoretical sampling perspective with, "the aim being to explore the dimensional

Table 2.2 Biosecurity threats and the frequency of articles reported

Biosecurity threat	Frequency	Percent of articles (%)
Dual use	66	50
Bioweapon	37	28
Ecological impact	29	22
Accidental release	26	20
Bioterrorism	23	17
Gain of function	16	12
Societal impact	16	12
Information access	12	9
Lower barriers	9	7
Uncertain consequences	7	5
DIY community	5	4
Difficult to monitor	4	3
Theft	4	3

Table 2.3 Biosecurity repsonses and the frequency of articles reported

Biosecurity response	Frequency	Percent of articles (%)
Legal oversight	59	44
Communication/open discussion	49	37
Risk assessment/management	45	34
Self-governance/ethics	37	28
Education/training/awareness	35	26
Collaboration	30	23
Biosafety principles	29	22
International governance/guidelines	25	19
Surveillance	19	14
Augment access	19	14
Improved response capacity	13	10
Containment or reversal strategies	7	5
Funding	6	5

range or varied conditions along which the properties of concepts vary" (Corbin and Strauss 1990, p. 73). Using the constant comparative method, each article was compared against one another to inductively assess potential emergent themes without *a priori* assumptions of the content or form of those themes. This inductive analysis design allows for sought-after themes to emerge from patterns present in the cases under analysis without presupposing what the important themes will be (Patton 2014). Next we report the key themes identified across our samples regarding biosecurity threats and responses.

2.3 Results and Discussion

2.3.1 US and EU Governmental Frameworks

2.3.1.1 US Framework

The process of creating a biodefense framework for the United States of America follows several engineering paradigms, with the Specify-Design-Build-Test-Learn (SDBTL) cycle being the guiding principle (NASEM 2018). In the SDBTL cycle, researchers are able to systemically identify the desired organism or the organism's functionality that will be genetically altered in a synthetic biology (SynBio) experiment. The researchers will then be able to assess the success of the alteration and amend the experimental protocol if needed. This empirical process has given fruition to a framework by the National Academy of Science, where certain parameters of biotechnology can be used to qualify a level of concern or hazard in the usability of a certain SynBio related technology. It must be noted that this framework aims not to enumerate the level of risk, but to direct concern to where the technology might be the most compromised. The concern about SynBio-related technology stems from the potential of its nefarious use, creating an objective assessment of achievements and shortcomings. The framework can be summarized into four parts: usability of technology, usability as a weapon, requirement of actors, and potential for mitigation. The four parts can be further broken down into subparts that can be assessed more easily.

Usability of Technology can be decomposed into four categories: ease of use, rate of development, barriers to use, and synergy with other technology. Ease of use is related to the commonality of the technology or of the information. The more widespread and accessible the technology is, the more accessible it is to nefarious actors, thereby increasing its threat. Rate of development refers to rapidity of improvements/innovations, whether there is a defined common use of the technology and if the technology is relevant throughout the times. For example, if a new technology does not have an intended use and there is a lot of funding behind the development of the product, the technology would generate concerns over its misuse. Barrier to use refers to the hurdles that can limit the use of technology, with hurdles being factors such as the accessibility of knowledge on how to operate the technology, the accessibility of materials needed for the technology, and other parameters. The lower the barrier of usage, the higher the concern about the technology. Synergy with other technology assesses whether there are other technologies present that can enhance the effects. Thus, the presence of synergy with other tools would increase the overall level of concern.

Usability as a Weapon is an assessment based on three other factors: production and delivery, scope of casualty, and predictability of results. Production and delivery refer to whether a genetically engineering organism, or any other product related to biotechnology, can produce toxins or other nefarious substances that can endanger people's lives. Production and delivery can be done inadvertently or carried out with a purpose. Therefore, as the production and delivery increases, the concern increases as well. Scope of casualty refers to the scale of the potential threat. A higher scope of casualty means more people are being endangered, which increases the level of concern. The predictability of results refers to the certainty of a nefarious user getting their intended results. This can be measured in the need for testing and if phenotype predictability is present. If testing is not needed to achieve the desired effects, then the level of concern and biosecurity threat increases since there are reduced opportunity for authorities to recognize and prepare for an attack for attack. If phenotype predictability is high, then the nefarious user has some confidence that the protocol they have followed will result in success. Therefore, as the phenotype predictability increases, so does the level of concern.

Requirements of Actors is an assessment of the feasibility of perpetrators successfully using specific biotechnology to commit a planned attack. The successful completion depends on the access to enterprise, access to resources, and organizational footprint. Access to enterprise relates to whether the actors have interaction with or access to the tools of question. Access to resources refers to whether the actors have the resources to carry out their attack. Resources can include items such as money, raw ingredients, and laboratory space. Lastly, organizational footprint is an estimate on how much manpower is needed to complete the attack. The more people an organization needs to complete an attack, the lower the concern level.

Potential for Mitigation is an assessment of areas of concern that can be addressed before an attack or an event occurs. The analysis is broken into four subparts to create a holistic examination of policies or accessibility issues that can compromise biosecurity. The first part is deterrence and prevention capabilities, which identify potentially dangerous activities and take steps towards preventing these activities. Actions such as increased intelligence gathering and instituting regulatory safeguards to areas of concern are effective in preventing certain tools from wreaking havoc. The second step in assuagement is developing the capability to recognize an attack. The identification process depends heavily on public health and disease databases as well as surveillance systems. By identifying and outlining the pre-existing tools available, the identification process can be optimized. Another step in mitigating a potential threat is attributing capabilities of an attack to a certain group. In simpler terms, matching the scientific evidence left at the attack to the organizations that have done it. The more difficult it is to identify the culprit, the higher the level of concern. The last step for mitigating a potential threat is developing an appropriate response to a myriad of attacks. Consequence management capabilities refers to a series of protocols and procedures that are established before an attack happens in order to quickly and efficiently respond to the attack and minimize the damage done. The procedures often include increasing emergency response capacity, developing quarantining facilities and expanding healthcare facilities.

2.3.1.2 EU Framework

The 2017 European Union (EU) Chemical, Biological, Radiological and Nuclear (CBRN) Action Plan outlines a general framework on improving prevention, preparedness, and response in case of an attack. The document also includes a clause that obligates EU member states to provide assistance to those victims of CBRN attacks and to maintain communication between countries within and outside the EU. Although the action plan covers responses to other threats that are not biological, most of the identified threats and responses are ubiquitous in application to biosecurity. The framework can be split into four actions: reducing accessibility of materials, ensuring preparedness for incidents, building stronger links, and enhancing current knowledge of risks.

The frameworks primary concern is **limiting the accessibility of potential dual-use technology or any other hazardous materials**. Increasing the legal control of law reinforcement and preventing the trade of dangerous material to nefarious actors is one viable option. Methods of accessibility reduction include providing technical reports on weapons and incidents through Europol and strengthening patrol at EU borders. This action can increase the awareness of potential threats in law enforcement personnel and hinder the spread of material that can be used to initiate an attack. Another implementation measure that can limit the accessibility of materials is to decrease insider threats by optimizing vetting and background checks of personnel in facilities holding CBRN materials in order to identify and remove individuals with nefarious intentions.

The second step identified in the action plan is to ensure that **member states are prepared to respond to a CBRN incident**. Due to the diversity of the European Union, safety protocols get adopted at various levels of intensity depending on the economic and political status of the member state. The first proposed action towards a uniform response to CBRN incidents is to develop a common training curriculum and institute cross-sectorial training and exercise. Other measures to improve overall preparedness for potential attacks include systematic review and assessments of previous CBRN Action Plans and strengthening the current European Emergency Response Capacity of the Union Civil Protection Mechanism by registering proposed CBRN modules. Updating current technologies and systems used for monitoring CBRN materials is imperative to stay relevant to the technologies available to nefarious actors. Conducting a gap analysis on CBRN material detection and improving the Early Warning and Response System (EWRS) can be a vital strategic advantage for EU states against malignant actors. Concurrently, improving the ability for laboratories to identify CBRN material and improving medical countermeasures such as joint efforts in research and development of vaccines should be pursued. Pertaining specifically to biosecurity, it is essential to increase overall awareness and develop a response protocol for emerging bio-risks.

Building stronger internal-external links in CBRN security with key regional and international EU partners was also identified in the framework. While most of the previous content was focused on strengthening the flow of information and resources within the EU, the later part of the framework focused on maintaining a similar level of contact with foreign entities outside of the EU. Particularly with NATO, the primary aim of the partnership is to develop a counter-terrorism protocol by increasing information exchange, capacity building, training, and exercise. This ties into the final section of the EU CBRN framework: **enhancing current knowledge on CBRN risks.** This last section focused primarily on the creation and application of a security network. The EU CBRN security network will be overseen by a dedicated advisory group, and will make information available for sharing with Europol. The maintenance of the research network will depend on updating prevalent needs and threats relating to CBRN.

2.3.2 Threats Identified in the Literature

The thirteen biosecurity threats identified in the literature are summarized in Table 2.2. On average, each article mentioned two threats (min = 0, max = 8). Each threat, as informed by the literature, is further described and discussed below.

Dual Use was the most frequently mentioned threat, appearing in half of the analyzed articles. Dual use research of concern is defined as "life sciences research that, based on current understanding, can be reasonably anticipated to provide knowledge, information, products, or technology that could be directly misapplied to pose a significant threat with broad potential consequences to public health and safety, agricultural crops and other plants, animals, the environment, material, or national security," (Lev and Samimian-Darash 2014). DiEuliis and Giordano (2018) further state that "any tool that imparts great capability also involves at least some risk, if not threat, that the power conferred by such capacity can be used to leverage or evoke a variety of ends." This is at the center of the concern over dual use research. The majority of biotechnology research and modernization is legitimate and done with the goal of benefiting society – that is, to beneficial ends. However, the same knowledge and techniques gained from beneficial research can be used maliciously. For instance, CRISPR-Cas9 is being used to perform targeted gene editing as a treatment for cancer, increasing our ability to treat cancer and reducing our reliance on toxic chemotherapy drugs, but it could also be used to edit pathogens to increase their virulence. Indeed, the dual use applications and threats from CRISPR are prominently featured in the literature (Vogel and Ouagrham-Gormley 2018; Webber et al. 2015).

Much of the controversial dual use research in biotechnology involves **gain of function** (GOF) studies, a term used in 12% of articles. Duprex et al. (2014) considers GOF to be a "generic label for a broad class of experiments that lead to a genetically altered biological agent with new or enhanced functions." Much of the concern over GOF studies includes research on the avian influenza virus and relatives of the smallpox virus (Evans et al. 2015; Duprex et al. 2014). These studies conferred new traits to the virus that increased its virulence in order to study transmission or vaccine creation, but also have a clear application for biological weapon (bioweapon) development. In this way, they are both GOF and dual use research. Publishing these sorts of studies is considered a biosecurity threat of its own because the information could allow a nefarious actor to create a bioweapon when they otherwise wouldn't have had the knowledge to do so. We refer to this threat as "**information access**" and it is mentioned in 9% of articles. DiEuliss and Gronvall (20) touch on this threat while writing about the controversial publication of a study that synthesized horsepox from scratch. They state that "horsepox is not a significant disease for humans, but there is concern that publication of these experiments could lower barriers toward the synthesis and booting up of another orthopoxvirus, variola (smallpox) virus, which was a significant scourge in history."

Bioweapons were the second most frequently mentioned threat, appearing in 28% of the articles analyzed. Franconi et al. (2018) define bioweapons as "deadly pathogens – bacteria or viruses – or toxins that can be deliberately released in order to cause harm to people or animals and plants." Generally, when a bioweapon is used by a state sponsored entity it is considered an act of biowarfare, while the use of a bioweapon by a non-state sponsored entity or individual is considered an act of **bioterrorism** (Jamil 2015), the latter of which was mentioned in 17% of articles. Unaltered organisms can and have been used as bioweapons in the past, such as in the 2001 anthrax attacks. Biotechnology opens the door to creating enhanced or novel pathogens and new avenues for toxin production. Cross (2018) identifies three ways in which biotechnology can be used to create bioweapons: (1) "recreating pathogenic viruses such as Ebola, SARS, or smallpox," (2) "engineering bacteria to make them more dangerous, which could be easily accomplished by inserting genes to confer antibiotic resistance," and (3) "engineering microbes to produce and release toxic biochemicals." Researchers have already demonstrated capabilities in all three of these avenues. Horsepox, a close relative of smallpox, has been synthesized from mail-ordered DNA (Noyce et al. 2018), avian influenza has been engineered to allow for airborne transmission between mammals (Linster et al. 2014), and botulinum toxin has been produced using yeast cells (Fonfria et al. 2018). These three cases are also prime examples of dual use research, as they were carried out for beneficial purposes (vaccine development, study of transmission, and enhanced therapeutics, respectively) but also provide a clear avenue towards weaponization.

Bioweapons and bioterrorism are mainly concerned with the deliberate release of an engineered pathogen or toxin with the purpose of causing harm, but the literature also identified **accidental releases** of modified organisms as a threat. Accidental releases are often cited as a concern for biosafety and not biosecurity. Pastorino et al. (2017) delineates the two terms in a laboratory setting as follows:

> "Laboratory biosafety" is the term used to describe the containment principles, technologies, and practices that are implemented to prevent unintentional exposure to pathogens and toxins or their accidental release. "Laboratory biosecurity" refers to institutional and personal security measures designed to prevent the loss, theft, misuse, diversion, or intentional release of pathogens and toxins."

Legitimate research on the most dangerous pathogens are often restricted to laboratories with a high Biosafety Level (BSL) designation – as regulated/monitored by a country's federal disease agency and the World Health Organization. These laboratories have the necessary precautions to drastically reduce the risk of inadvertently releasing wild type or engineered pathogens (Imperiale et al. 2018). Accidental releases are a higher threat if work is being done in laboratories without proper safety measures, such is in the growing biotechnology Do-It-Yourself (DIY) community. The risk of unintentional releases also increases as engineered organisms are taken outside of the laboratory setting, such as in the case of bacteria engineered for soil bioremediation or algae engineered for biofuel production and grown in outdoor open-air tanks (Mandel and Marchant 2014). In both of these cases, the engineered organisms have the potential to escape outside of the intended soil or water and have unknown consequences for the receiving environment.

Regardless of how an engineered organism or virus makes it out into the open, the potential ecological and societal impacts were frequently identified as concerns. Interestingly, concerns over ecological impacts appeared in 22% of articles, almost twice as many articles as societal impacts which appeared in 12% of articles. While bioweapons could be created to directly attack some critical component the environment, the concern over **ecological impacts** is largely driven by the increased interest in using engineered organisms for controlling nuisance species and recent advancements in gene drive technology. Weidmann (2018) defines gene drives as the "experimental techniques which are supposed to push foreign genes into the chromosomes of wild populations with the aim to change the complete organisms in just a few generations." Popular examples of the potentially beneficial application of gene drives include reducing populations of mosquitos responsible for spreading dengue fever and malaria (Finkel et al. 2019; Weidmann 2018) and exterminating introduced rodents that predate on endangered reptiles and birds from islands (Weidmann 2018). However, since biological organisms and viruses are capable of reproducing, mutating, and sharing genes, there is little way to guarantee that the intended modification will be contained in the target population only and little way to anticipate the cascading environmental consequences of manipulating populations of species in such a way. This is at the heart of the threat advanced biotechnology poses to ecological systems and is discussed in multiple papers. Weidmann considers gene drives "ethically questionable because we still do not know if the genetic changes could affect other organisms or even entire ecosystems in a negative way." Wintle et al. (2017) raises the similar concern that "deploying gene drives in wild populations might alter ecosystems, disrupting trophic levels and food webs, and creating vacant niches (for example, for new disease vector species or new disease organisms)." Webber et al. (2015) conclude that "removing species with gene drive technology could produce unintended cascades that may represent a greater net threat than that of the target species." Overall, concerns of the negative and irreversible impact that one genetically engineered species could have on the entire ecosystem was evident in the literature.

These quotes, and the inherent ability of biological systems to mutate and evolve, also demonstrate the threat of the **uncertain consequences** of utilizing and deploying advanced biotechnology, which was expressed in 5% of articles. Ecological systems are filled with complex, intricate, and unknown interactions (from the global to the intracellular level) and advanced biotechnology is still a young and rapidly developing field with only a few examples of field trials with engineered organisms (Seager et al. 2017). When these two complex components are considered in tandem, it is not surprising to see in the literature that some experts are concerned that consequences cannot be reliably foreseen and avoided.

Threats to human health are abundant in the conversation of bioweapons and bioterrorism, in which enhanced pathogens that could cause mass human casualty are a primary fear. However, **societal impacts** as identified in this chapter are less concerned with human health and more concerned with human or environmental modification and how these alterations would impact society. Caballero-Hernandez et al. (2017) suggest that gene drives used to control nuisance species could impact a nation's food security and national sovereignty, though they do not elaborate on how. Esvelt and Gemmell (2017) also mention the problem of national sovereignty and gene drives, in that one nation risks infringing on the national sovereignty and harming diplomatic relations with another nation if they release an engineered strain of a species that is found in both nations without the other nation's consent since the engineered strain will cross national borders. Concerning human modification, genome editing has been proposed as a way to remove undesirable traits from a human population, increase the average cognitive ability of a nation, and enhance combat soldiers by decreasing the need for sleep and food (Esvelt and Millett 2017), all of which have serious equality, security, and societal implications. Gomez-Tatay et al. (2016) propose that synthetic biology could be used to "improve humans and to develop what it has been called sub-humans, a kind of humanoid organism which would serve several purposes, such as being sources of transplantable tissues and organs, experimental subjects or crash test dummies and landmine diffusers." While this vision of sub-humans shows potential in improving life and safety for modern humans, it also has clearly negative ethical and societal implications. Considering human modification, the field of biotechnology would need to make leaps and bounds forward in order for these threats to be realized, but nonetheless they are important to consider and address as biotechnology progresses and advanced engineering of humans becomes more possible.

The **lowering of barriers** to entry into the biosecurity field and the **DIY biotechnology community** were identified as threats in 7% and 4% of articles, respectively. The lowering of barriers is largely caused by the increased globalization and democratization of the field in the past decade that has greatly increased the accessibility of the field to a wider number and diversity of people. While this has driven innovation and resulted in many beneficial applications, it has also reduced the barriers that would have previously kept nefarious individuals – whether working alone or for an organized state, sub-state, or non-state group – from using biotechnology towards their own harmful ends. It has also allowed individuals in the DIY community to construct quasi-laboratories in their own homes and carry out their own experiments devoid of regulations or safety precautions. DiEuliis and Giordano (2018) highlight these threats in relation to gene editing by stating that "the relative availability of [gene editing techniques] enables increasing use by public research and do-it-yourself (i.e., biohacking) communities which could foster risk incurred by both inadvertent misuse and/or intentional development of products that threaten public safety." Jefferson et al. (2014) share this sentiment, but extend it beyond the DIY community by expressing fears that "the 'de-skilling' of biology, combined with online access to the genomic DNA sequences of pathogenic organisms and the reduction in price for DNA synthesis, will make biology increasingly accessible to people operating outside well-equipped professional research laboratories, including people with malevolent intentions." The ease with which people are able to access information, equipment, materials, and learn techniques is therefore a growing biosecurity threat.

The final two biosecurity threats identified in the literature were the **difficulty of monitoring** and potential **theft of pathogens or equipment**, which both appeared in 3% of articles. A number of concerns fall under the category of difficult to monitor. These include the difficulty of determining which organisms/viruses and genetic modifications could be used maliciously and monitoring for them (DiEuliis and Giordano 2017), of monitoring the spread of an engineered trait beyond where it was intentionally deployed for species control (Webber et al. 2015), and of differentiating a natural outbreak from a biological attack (MacIntyre et al. 2018; Nelson et al. 2014). Regarding theft, MacIntyre et al. (2018) and Walsh (2016) both see the ability for radicalized research staff to steal pathogens from the laboratories they have access to as a biosecurity threat. Berger and Schneck (2019) and Kozminski (2015) are additionally concerned over the threat of malicious actors stealing sensitive data that is stored digitally. Kozminski (2015) cautions that "in the area of Big Data with specific applications to the life sciences, information taken could potentially be used for exploitation or extortion." This "Big Data" includes the ever-growing databases devoted to people's genetic information collected for forensic, genealogical, or research purposes.

2.3.3 Responses Identified in the Literature

The thirteen biosecurity responses identified in the literature are summarized in Table 2.3. On average, each article mentioned three threats (min = 0, max = 8). Each response, as informed by the literature, is further described and discussed below.

Legal Oversight or Regulations at the national level were the most frequently mentioned biosecurity response, appearing in just under half of the articles analyzed. However, the form and extent of that regulation varied. The US Government Policy for the Oversight of Life Sciences Dual Use Research of Concern places restrictions on certain types of experiments on certain infectious agents and toxins (Lev and Samimian-Darash 2014). De Beer and Jain (2018) suggest that regulations need to remain loose enough to allow for innovation and that outreach and monitoring can supplement such loosening in oversight. Some articles call for regulations throughout the research and development process (Gomez-Tatay and Hernandez-Andreu 2019), carefully scrutinizing the primary investigator, purpose, location, and process, while others advocate that mainly the end product should be subject to regulation (Gronvall 2015). Regulations dictating who can purchase what genetic material and equipment, which laboratories are approved to conduct dual use synthetic biology research, and what knowledge is appropriate to disseminate in journals were also suggested in multiple articles (Gomez-Tatay and Hernandez-Andreu 2019; Pope 2017; Adam et al. 2017; Diggans and Leproust 2019; Marris et al. 2014). These regulations also fall into **augmentation of access**, which appeared in 14% of articles and could be considered a sub-category of legal oversight. The augmentation of access includes any measures that reduce a person's ability to access equipment, materials, facilities, or knowledge required to partake in synthetic biology, thereby reducing the risk of unauthorized personnel engaging in intentionally or unintentionally harmful research.

International Agreements, Guidelines, or Regulations were also suggested, though at 19% of papers this response appeared less than half as often as national regulations did. The driving thought behind some form of international governance, in addition to the national regulations discussed above, is that any accident or attack with an engineered organism is likely to have far-reaching consequences for an entire ecoregion, continent, or the world. It is therefore in the best interest of humanity for all nations to come together and agree on best practices as they relate to advanced biotechnology. The BWC, as discussed in the introduction, is a foundational international agreement in which nations agree not to create or stockpile bioweapons. Bioweapons can be easily and objectively viewed as "bad," making regulations against them relatively simple, but much of modern advanced biotechnology exists in a more complicated grey zone owing to its dual use potential and newness as a field (Greer and Trump 2019). Experts have called for new international agreements as the field of biotechnology has diversified and its related threats have expanded past just biological weapons. A prime example is the risk posed by releasing gene drives into the environment. Redford et al. (2014) emphasize that this poses a relatively new threat and that "international regulation of the development and release of modified organisms needs considerable work," and will require "wider competence on the part of diplomats and lawyers in understanding both synthetic biology and ecology." Other experts suggest that regulations on advanced biotechnology could be applied under existing international treaties and agreements, such as the Convention on Biological Diversity, Nagoya Protocol, United Nations Security Council Resolution 1540 (non-proliferation of weapons of mass destruction), and the BWC (Gronvall 2015; Stewart 2018; Ahteensuu 2017). Citing new rules or guidelines under existing agreements that nations have already agreed to is viewed as a more stream-lined method than creating entirely new treaties and agreements.

The third most frequently mentioned response is **risk assessment or management**, which appeared in 34% of the articles we analyzed. The most frequently mentioned risk assessment method was the risk-benefit analysis, in which "the risks of potential misuse (accidental or intentional) are weighed against the assumed potential benefits of scientific innovation," (Jacobsen et al. 2014). This analysis and other risk assessment methodology can be used to determine if a proposed study should occur. If the benefits outweigh the risks and the study is given the okay, then risk management can be used to identify "how to do it safely and mitigate risks," (Imperiale and Casadevall 2018). Risk management options proposed in the literature include laboratory biosafety (Pastorino et al. 2017), containment strategies (Duprex et al. 2014), publication restrictions (Rychnovska 2016), and more. Multiple authors called for risk assessment and subsequent plans for risk mitigation be conducted during the project planning/grant application phase. Oeschger and Jenal (2018) argue that successful risk assessment and management requires the input of "the life science research community itself as proper risk evaluation and management depends on expert knowledge." Suk et al. (2014) add that risk assessment "needs to integrate the best available information from a variety of sectors, meaning that life scientists, regulators, ethicists, public health actors, and the security and intelligence communities will need to become more adept at and comfortable with exchanging information and ideas." In this quote, Suk et al. (2014) also demonstrate the benefit of **collaboration** between experts from diverse fields in reducing biosecurity threats. The usefulness of collaboration as a response was identified in 23% of articles, and was suggested as a way to improve the identification of an outbreak (MacIntyre 2015), policy design and implementation (Edwards 2014), laboratory biosafety (Trevan 2015), public outreach (Redford et al. 2014), and biological data security (Berger and Schneck 2019), in addition to risk assessment.

Three out of the five most frequently mentioned responses had less to do with government oversight and more to do with social aspects: **communication and open discussion** between scientists, government, industry, and the public (37%); a strong sense of **ethics and self-regulation** amongst scientists (28%); and proper **training of scientists and awareness of biosecurity concerns** (27%). These three social responses are also complimentary to one another. Oeschger and Jenal (2018) state that "a code of conduct intends to promote ethical principles and corresponding behavioral norms that often go beyond legal requirements." By adopting a code of conduct, scientists "raise awareness of and foster responsibility for dual use aspects of life science research within the scientific community," (Oeschger and Jenal 2018) Fear and ter Meulen (2016) further define self-regulation as a system in which "there are checks and balances within the scientific community, not [one in which] each researcher is free to decide unilaterally which procedure to follow." Self-regulation not only requires open communication between scientists, but also for scientists to have a line of communication with the public, government regulators, and other stakeholders in order to be aware of the concerns surrounding biotechnology, see how their intended research relates to those concerns, and take appropriate actions to respond. Scientists should also communicate with the public to relieve unwarranted concerns held by the public and allow scientific research to continue. Baskin (2019) emphasizes that "intentional, careful, and reassuring communications from the scientific community to the public benefit both science and the public." Baskin (2019) stresses that while self-regulation is ideal, scientists should also receive training in the law-making process and how to engage with it so that the scientific community can "become involved throughout the rule-making process to prevent excessive restrictions that are potentially counter-productive to national biosecurity" when legal instruments are unavoidable. According to the literature, training, communication, and self-governance have great potential for addressing biosecurity threats (Engel-Glatter and Ienca 2018; Oeschger and Jenal 2018; Baskin 2019; Gomez-Tatay et al. 2016).

The use of **biosafety principles** to reduce biosecurity threats, particularly accidental releases of engineered organisms, was identified in 22% of articles. Biosafety principles include the design of laboratories with precautions appropriate to the risk-level of pathogens being studied (BSL designations, as mentioned above), "train[ing] people that work there, the implementation of regulations, and the use of robust risk-based approaches to mitigate adverse events," (Vogel et al. 2015). Fear and ter Meulen (2016) emphasize that "attention to key biosafety issues is imperative at all stages of the research endeavor from first formulating a research idea through to the publication of results." By identifying and following appropriate biosafety precautions, studies with advanced biotechnology can be conducted with the confidence that accidental releases will not occur, that the general public and local environment will not be affected, and that workers are properly protected while performing their duties. Certain biosafety principles, such as the requirements to meet different BSL classifications, are regulated by state or federal agencies, but additional requirements could be established by individual institutions.

Many of the responses we identified looked to reduce the risk of a biosecurity threat before it could occur, but it is also important to have the capacity to recognize and respond to threats once they are present. Building a nation's **capacity to respond** to a biosecurity threat, also referred to as "preparedness strategies," before the threat is present and actively engaging in **surveillance** of present or imminent biosecurity threats were identified in 10% and 14% of articles, respectively. Nelson et al. (2014) summarizes the impact that both of these responses can have: "Surveillance strategies enable early detection, which is vital for rapid and effective emergency responses whilst preparedness strategies are essential for maintaining a nation's capacity to carry out effective response and recovery processes." They go on to report three types of surveillance that Australia uses to identify unusual disease patterns that could indicate an outbreak: "passive surveillance, involving routine reporting of certain disease cases; active surveillance, involving the specific collection of data relating to a particular disease; and sentinel surveillance, where data are collected from a subpopulation to provide an indication of trends in the wider population," (Nelson et al. 2014). Surveillance of DNA sequence orders made to DNA synthesis companies to identify and terminate potentially malicious orders has been practiced and suggested for wider adoption, as has monitoring social media and the dark web for signs that a biological attack is being planned or has occurred (MacIntyre et al. 2018). Improved response capacity includes a variety of measures aimed at quickly recovering from a biological attack, thereby reducing the amount of damage that can be done. According to Nelson et al. (2014), "preparedness strategies incorporate aspects including: planning; personnel training; monitoring and reviewing policies and programs; maintaining supply stocks; and carrying out ongoing research into improved methods for disease diagnosis, treatment or prevention." Preparedness strategies proposed in the literature include the stockpiling of vaccines and personal protective equipment (Adam et al. 2017), restructuring public health organizations and training medical personnel to react to a biological attack (de Almeida 2015), and creating novel platforms for rapid disease diagnostics and vaccine production (Franconi et al. 2018).

One response mentioned mostly in articles concerned with the threat of gene drives and their potential ecological impact was the creation and use of **containment and reversal strategies**, which we identified in 5% of articles. These strategies have been referred to as "risk-reducing innovation" and "built-in safety," and include creating strains of a modified organism that cannot survive outside of the laboratory or can only live on specific substrates to eliminate the threat of accidental release (van de Poel and Robaey 2017), modifying existing genes in a way that allows for the ancestral sequence to be easily restored (i.e. genetic restoration) (Looi et al. 2018), adding susceptibilities to specific treatments or chemicals (i.e., kill switches) that would allow the engineered population to be easily controlled (Wintle et al. 2017), and only building and testing gene drives in geographic areas where the target species is not naturally present (Esvelt et al. 2014). These strategies work to reduce the threat of accidental releases, ecological impacts, and uncertain consequences right at the beginning of the study by building in a way to restore the engineered organism back to its natural state or eliminate it completely.

The final and least frequently mentioned response was **funding**, which appeared in 5% of articles. These articles called for the funding of specific threads of research or institutions that would help to enhance biosecurity. Evans (2014) stated that large "funding bodies have a key role to play reshaping our understanding of what it means to engage in biosecurity governance," and believes that studies looking directly at the social aspects of emerging biotechnology should be funded in their own right and not just as add-ons to other research as has been done previously. He believes that these sorts of studies will allow governance to progress alongside the biotechnology field instead of playing catch-up, but have lacked funding to date. Other calls for funding to increase biosecurity include ensuring that laboratories are financially able to implement proper biosafety precautions (Trevan 2015), offsetting costs for DNA synthesis companies to screen for orders related to pathogens (DiEuliis et al. 2017), and funding biotechnology companies to increase innovation within the nation and decrease the likelihood of these companies moving oversees (Gronvall 2015).

2.4 Conclusion

Biosecurity threats and responses have garnered significant attention and these issues warrant continued investigation and prioritization in order to maximize the benefits and reduce the likelihood of misuse that could cause significant harm to human and environmental health. This review pulls disparate literature together to provide description of the field in sum to date. We envision scholars and decision-makers to use this work to forward new research agendas to better allocate resources toward underdeveloped, yet valuable areas with prescient needs.

By conducting a systematic literature review, we were able to determine which biosecurity threats and responses are most prevalent across this broad field. Dual use research was far and away the most frequently mentioned threat, appearing 29 more times than the second most frequent threat biological weapons. Both of these threats were also prominently evident in the US and EU frameworks developed to address biosecurity concerns evolving out of the use of advanced biotechnology. The US framework also identified information access and lower barriers as biosecurity threats, while the EU framework mentioned the threat of theft by staff ("insider threat") (Trump et al. 2020c). Overall, the US and EU frameworks were concerned with preventing and responding to attacks with biological weapons. While this was a key threat identified in the literature, many of the other threats discussed were more concerned with the potential for negative consequences of authorized releases of engineered organisms into the environment. This accounts for the third most frequently mentioned threat in the literature, ecological impact, which was absent from the governmental frameworks.

Legal oversight was the top response identified in the literature and was also presented as a biosecurity response in both the US and EU frameworks. These frameworks also included other government-driven responses identified in the literature, including surveillance, augmentation of access, improved response capacity, and risk assessment. The EU framework additionally suggested international governance, collaboration, and training/awareness as responses, but these were referring to collaboration and training of governmental agencies (Trump 2017). The top responses identified in the literature that applied more to industry, academia, and the public (open discussions, self-governance and ethics, and education/training/awareness) were noticeably missing from the two governmental frameworks included in this chapter.

This empirical foundation of the prominent areas of concern for biotechnology-related research and discourse may be used to help formulate needs-based considerations for future research. Concise understanding and acknowledgement of the spectrum of concerns related to the proliferation of biotechnological tools can inform regulators and decision-makers who must hold command over contemporary concerns and this work should be used to enable better informed decisions about priority tasks to corral biosecurity threat and foster adaptive responses. As this area continues to gain prominence within communities concerned with biotechnological risk, we anticipate the topics covered here to grow in coverage at an increasing rate and we anticipate the entrance of yet determined considerations for novel threats and responses. Thus, in due time, we feel a replication of this method and results is warranted to isolate threat and response developments post-2020.

References

Adam DC, Magee D, Bui CM, Scotch M, MacIntyre CR (2017) Does influenza pandemic preparedness and mitigation require gain-of-function research? Influenza Other Respir Viruses 11(4):306–310

Ahteensuu M (2017) Synthetic biology, genome editing, and the risk of bioterrorism. Sci Eng Ethics 23:1531–1561

Baskin CR (2019) Who should be driving US science policy? Perspect Biol Med 62(1):20–30

Bauer MW, Gaskell G (eds) (2002) Biotechnology: the making of a global controversy. Cambridge University Press, Cambridge

Berger KM, Schneck PA (2019) National and transnational security implications of asymmetric access to and use of biological data. Front Bioeng Biotechnol 7:21

Caballero-Hernandez D, Rodriguez-Padilla C, Lozano-Muniz S (2017) Bioethics for biotechnologists: from Dolly to CRISPR. Open Agric 2(1):160–165

Corbin JM, Strauss A (1990) Grounded theory research: procedures, canons, and evaluative criteria. Qual Sociol 13(1):3–21

Cross R (2018) Synthetic biology poses new biosecurity risks. Chem Eng News 16

De Almeida ME (2015) The permanent relation between biology, power and war: the dual use of the biotechnological development. Ciencia & Saude Coletiva 20(7):2255–2266

De Beer J, Jain V (2018) Inclusive innovation in biohacker spaces: the role of systems and networks. Technol Innov Manag Rev 8(2):27–37

DiEuliis D, Giordano J (2017) Why gene editors like CRISPR/Cas may be a game-changer for neuroweapons. Health Secur 15(3):296–302

DiEuliis D, Giordano J (2018) Gene editing using CRISPR/Cas9: implications for dual-use and biosecurity. Protein Cell 9(3):239–240

DiEuliis D, Carter SR, Gronvall GK (2017) Options for synthetic DNA order screening, revisited. MSphere 2:e00319–e00317

Diggans J, Leproust E (2019) Next steps for access to safe, secure DNA synthesis. Front Bioeng Biotechnol 7:86

Duprex WP, Fouchier RAM, Imperiale MJ, Lipsitch M, Relman DA (2014) Gain-of-function experiments: time for a real debate. Nat Rev Microbiol 13:58–64

Edwards B (2014) Taking stock of security concerns related to synthetic biology in an age of responsible innovation. Front Public Health 2:79

Engel-Glatter S, Ienca M (2018) Life scientists' views and perspectives on the regulations of dual-use research of concern. Sci Public Policy 45(1):92–102

Esvelt KM, Gemmell NJ (2017) Conservation demands safe gene drive. PLoS Biol 15(11):e20003850

Esvelt KM, Millett PD (2017) Genome editing as a national security threat. Rev Sci Tech 36(2):459–465

Esvelt KM, Smidler AL, Catteruccia F, Church GM (2014) Concerning RNA-guided gene drives for the alteration of wild populations. Elife 3:e03401

European Commission (2017) Action Plan to enhance preparedness against chemical, biological, radiological and nuclear security risks. Brussels, Belgium

Evans SG (2014) What's the matter with biosecurity? J Responsib Innov 2(1):88–91

Evans NG, Lipsitch M, Levinson M (2015) The ethics of biosafety considerations in gain-of-function research resulting in the creation of potential pandemic pathogens. J Med Ethics 41(11):901–908

Fear R, ter Meulen V (2016) European academies advise on gain-of-function studies in influenza virus research. J Virol 90(5):2162–2164

Finkel AM, Trump BD, Bowman D, Maynard A (2019) A "solution-focused" comparative risk assessment of conventional and synthetic biology approaches to control mosquitoes carrying the dengue fever virus. Environ Syst Decis 38:177–197

Fonfria E, Elliott M, Beard M, Chaddock JA, Krupp J (2018) Engineering botulinum toxins to improve and expand targeting and SNARE cleavage activity. Toxins 10(7):278

Franconi R, Illiano E, Paolini R, Venuti A, Demurtas OC (2018) Rapid and low-cost tools derived from plants to face emerging/re-emerging infectious diseases and bioterrorism agents. In: Radosavljevic V, Banjari I, Belojevic G (eds) Defense against bioterrorism. Springer, New York

Gomez-Tatay L, Hernandez-Andreu JM (2019) Biosafety and biosecurity in synthetic biology: a review. Crit Rev Environ Sci Technol 49(17):1587–1621

Gomez-Tatay L, Hernandez-Andreu JM, Azner J (2016) A personalist ontological approach to synthetic biology. Bioethics 30(6):397–406

Greer SL, Trump B (2019) Regulation and regime: the comparative politics of adaptive regulation in synthetic biology. Policy Sci 52(4):505–524

Gronvall GK (2015) US competitiveness in synthetic biology. Health Secur 13(6):378–389

Hood L, Rowen L (2013) The human genome project: big science transforms biology and medicine. Genome Med 5:79

Imperiale MJ, Casadevall A (2018) A new approach to evaluating the risk–benefit equation for dual-use and gain-of-function research of concern. Front Bioeng Biotechnol 6:21

Imperiale MJ, Howard D, Casadevall A (2018) The silver lining in gain-of-function experiments with pathogens of pandemic potential. Methods Mol Biol 1836:575–587

Jacobsen KX, Mattison K, Heisz M, Fry S (2014) Biosecurity in emerging life sciences technologies, a Canadian public health perspective. Front Public Health 2:198

Jamil SAB (2015) Ethics in synthetic biology: exacerbated misconceptions of the nature of man and cosmology. Asian Bioethic Rev 7(3):331–337

Jefferson C, Lentzos F, Marris C (2014) Synthetic biology and biosecurity: challenging the "myths". Front Public Health 2:115. Front Public Health 2:198

Kozminski KG (2015) Biosecurity in the age of Big Data: a conversation with the FBI. Mol Biol Cell 26(22):3894–3897

Lev O, Samimian-Darash L (2014) Biosecurity policy in the US: a critical assessment. Front Public Health 2:110

Linster M, Boheeman S, de Graaf M, Schrauwen EJA, Lexmond P, Manz B, Bestebroer TM, Baumann J, van Riel D, Rimmelzwaan GF, Osterhaus ADME, Matrosovich M, Fouchier RAM, Herfst S (2014) Identification, characterization, and natural selection of mutations driving airborne transmission of A/H5N1 virus. Cell 157(2):329–339

Looi FY, Baker M, Townson T, Richard M, Novak B, Doran TJ, Short KR (2018) Creating disease resistant chickens: a viable solution to avian influenza? Viruses 10(10):561

MacIntyre CR (2015) Biopreparedness in the age of genetically engineered pathogens and open access science: an urgent need for a paradigm shift. Mil Med 180(9):943–949

MacIntyre CR, Engells TE, Scotch M, Heslop DJ, Gumel AB, Poste G, Chen X, Herche W, Steinhofel K, Lim S, Broom A (2018) Converging and emerging threats to health security. Environ Syst Decis 38:198–207

Mandel GN, Marchant GE (2014) The living regulatory challenges of synthetic biology. Iowa Law Rev 100(1):155–200

Marris C, Jefferson C, Lentzos F (2014) Negotiating the dynamics of uncomfortable knowledge: the case of dual use and synthetic biology. BioSocieties 9:393–420

National Academies of Sciences, Engineering, and Medicine (2018) Biodefense in the age of synthetic biology. The National Academies Press, Washington, DC

Nelson M, Roffey P, McNevin D, Lennard C, Gahan ME (2014) An overview of biosecurity in Australia. Aust J Forensic Sci 46(4):383–396

Noyce RS, Lederman S, Evans DH (2018) Construction of an infectious horsepox virus vaccine from chemically synthesized DNA fragments. PLoS One 13(1):e0188453

Nuclear Threat Initiative (NTI). The Biological Threat (2015). https://www.nti.org/learn/biological/

Oeschger FM, Jenal U (2018) Addressing the misuse potential of life science research – perspectives from a bottom-up initiative in Switzerland. Front Bioeng Biotechnol 6:38

Pastorino B, de Lamballerie X, Cherrel R (2017) Biosafety and biosecurity in European containment level 3 laboratories: focus on French recent progress and essential requirements. Front Public Health 31(5):121

Patton MQ (2014) Qualitative research & evaluation methods: integrating theory and practice. Sage, Thousand Oaks

Polaris Marketing Team (2020) Biotechnology market size, share, trends & analysis report 2020–2026. https://www.polarismarketresearch.com/industry-analysis/biotechnology-market. Accessed 20 Sept 2020

Pope SM (2017) Impact of gene editing tools, like CRISPR/Cas9, on the public health response to disease outbreaks. Disaster Med Public Health Prep 11(2):155–159

Raposo VL (2019) The first Chinese edited babies: a leap of faith in science. JBRA Assist Reprod 23(3):197–199

Redford KH, Adams W, Carslon R, Mace GM, Ceccarelli B (2014) Synthetic biology and the conservation of biodiversity. Oryx 48(3):330–336

Rychnovska D (2016) Governing dual-use knowledge: from the politics of responsible science to the ethicalization of security. Secur Dialogue 47(4):310–328

Sahl JW, Pearson T, Okinaka R et al (2016) A Bacillus anthracis genome sequence from the Sverdlovsk 1979 autopsy specimens. MBio 7(5):e01501-16. Published 27 September 2016. https://doi.org/10.1128/mBio.01501-16

Seager TP, Trump BD, Poinsatte-Jones K, Linkov I (2017) Why life cycle assessment does not work for synthetic biology. Environ Sci Technol 51(11):5861–5862

Stewart IJ (2018) Preventing weapons of mass destruction proliferation: the future of UNSCR 1540. In: Salisbury D, Stewart IJ, Viski A (eds) Preventing the proliferation of WMDs: measuring the success of UN Security Council resolution 1540. Palgrave Pivot, London, pp 105–126

Suk JE, Bartels C, Broberg E, Struelens MJ, Ozin AJ (2014) Dual-use research debates and public health: better integration would do no harm. Front Public Health 2:114

Trevan T (2015) Biological research: rethink biosafety. Nature 527(7577):155–158

Trump BD (2017) Synthetic biology regulation and governance: lessons from TAPIC for the United States, European Union, and Singapore. Health Policy 121(11):1139–1146

Trump BD, Cummings CL, Kuzma J, Linkov I (2020a) Synthetic biology 2020: frontiers in risk analysis and governance. Springer, Cham

Trump BD, Galaitsi SE, Appleton E, Bleijs DA, Florin MV, Gollihar JD et al (2020b) Building biosecurity for synthetic biology. Mol Syst Biol 16(7):e9723

Trump BD, Keisler JM, Volk KM, Linkov I (2020c) Biosecurity demands resilience. Environ Sci Technol

UNODA (2017) Biological weapons. https://www.un.org/disarmament/wmd/bio/. Accessed 21 Sept 2020

USDA APHIS (United States Department of Agriculture Animal and Plant Health Inspection Service (2020) Animal disease information. https://www.aphis.usda.gov/aphis/ourfocus/animalhealth/animal-disease-information. Accessed 14 Sept 2020

Van de Poel I, Robaey Z (2017) Safe-by-design: from safety to responsibility. NanoEthics 11:297–306

Verma AS, Agrahari S, Rastogi S, Singh A (2011) Biotechnology in the realm of history. J Pharm Bioallied Sci 3(3):321–323

Vogel KM, Ouagrham-Gormley SB (2018) Anticipating emerging biotechnology threats: a case study of CRISPR. Politics Life Sci 37(2):203–219

Vogel KM, Ozin AJ, Suk JE (2015) Biosecurity and dual-use research: gaining function – but at what cost? Front Public Health 3:13

Walsh PF (2016) Managing emerging health security threats since 9/11: the role of intelligence. Int J Intell Counterintell 29(2):341–367

Webber BL, Raghuc S, Edwards OR (2015) Opinion: is CRISPR-based gene drive a biocontrol silver bullet or global conservation threat? PNAS 112(34):10565–10567

Weidmann AG (2018) Frontiers in CRISPR. ACS Chem Biol 13(2):296–304

Wells E, Trump BD, Finkel AM, Linkov I (2020) A solution-focused comparative risk assessment of conventional and emerging synthetic biology technologies for fuel ethanol. In: Synthetic biology 2020: frontiers in risk analysis and governance. Springer, Cham, pp 223–255

Wintle BC, Boehm CR, Rhodes C (2017) A transatlantic perspective on 20 emerging issues in biological engineering. eLife 14(6):e30247

3

Biosecurity and Synthetic Biology Top-Down Governance Measures

R. Alexander Hamilton, Ruth Mampuys, S. E. Galaitsi, Aengus Collins, Ivan Istomin, Marko Ahteensuu and Lela Bakanidze

3.1 Introduction

Synthetic biology promises to make biology easier to engineer (Endy 2005), enabling more people in less formal research settings to participate in modern biology. Leveraging advances in DNA sequencing and synthesis technologies, genetic assembly methods based on standard biological parts (e.g. BioBricks), and increasingly precise gene-editing tools (e.g. CRISPR), synthetic biology is helping increase the reliability of and accessibility to genetic engineering. Although potentially enabling tremendous opportunities for the advancement of the global bioeconomy, opening new avenues for the creation of health, wealth and environmental sustainability, the possibility of a more 'democratic' (widely accessible) bioengineering capability could equally yield new opportunities for accidental, unintended or deliberate misuse. Consequently, synthetic biology represents a quintessential 'dual-use'

R. A. Hamilton
United Nations Interregional Crime and Justice Research Institute (UNICRI), Turin, Italy

R. Mampuys
Netherlands Commission on Genetic Modification (COGEM), Bilthoven, The Netherlands

S. E. Galaitsi (✉)
US Army Corps of Engineers, Concord, MA, United States

A. Collins
EPFL International Risk Governance Center, Lausanne, Switzerland

I. Istomin
Schaffhausen Institute of Technology (SIT), Schaffhausen, Switzerland

M. Ahteensuu
University of Turku, Turku, Finland

L. Bakanidze
EU CBRN CoE Regional Secretariat for Central Asia, Tashkent, Uzbekistan

biotechnology – a technology with the capacity to enable significant benefits and risks (NRC 2004).

In relation to existing top-down governance[1] measures for biosecurity,[2] synthetic biology represents a promising yet potentially destabilizing advancement in the life sciences, one that could introduce new risks and regulatory challenges. In particular, a number of high-profile synthetic biology experiments, ranging from the *de novo* synthesis of poliovirus (Cello et al. 2002) to the recent synthesis of horsepox virus (Noyce et al. 2018), have raised concerns that the same techniques could be exploited to bypass regulatory controls (e.g. the United States, US, Select Agent Regulations) on lists of high-risk pathogens. Moreover, the possibility of synthesizing novel 'taxonomically unclassified' pathogens (NSABB 2006; Garfinkel et al. 2007) has led some to question the logic and utility of current 'list-based' approaches to regulation. Looking to the future, if synthetic biology does, in fact, 'deskill' the 'art' of biological engineering, new regulatory approaches could very well be essential because the tools of modern biology will be widely accessible to both responsible and malicious actors.

Claims about synthetic biology's potential, like other emerging technologies, nonetheless tend to overstate its 'enabling' capacity. Likewise, the ease of producing biological weapons tends to be overstated. As a number of commentators note, biology is not yet easy to engineer (Jefferson et al. 2014) and, for the foreseeable future, the skills necessary to produce biological weapons are likely to remain only within the grasp of states (Piers Millet in Regalado 2016). However, the field's emphasis on eliminating technical barriers and reducing the importance of tacit knowledge (Oye 2012) represents a powerful source of expectation for advocates and critics alike. For advocates, it represents the possible realization of modern biology's full potential, one that could yield revolutionary advances in health, medicine, and industry in the twenty-first century. For critics, it represents a seemingly open-ended risk that requires exceptional precaution. For national governments, and international conventions responsible for establishing global biosecurity norms and obligations that are operationalized at the national level through legislation and other regulatory tools (McLeish and Nightingale 2007), a central question is how (if at all) does top-down biosecurity governance need to change in response to synthetic biology?

Regulatory considerations of this kind are both familiar and new (Hamilton 2015). In the 1970s, recombinant DNA technology similarly emerged as a source of significant and contrasting expectations, and questions were posed about the

---

[1] In this chapter, 'top-down governance' is taken to mean laws, regulations, policies, guidelines and other government-led regulatory measures aimed at prohibiting undesirable behavior or encouraging desirable behavior on the part of countries, organizations or individuals engaged in aspects of the life sciences (research activities involving the use of biological materials, knowledge and/or technologies).

[2] In this chapter, 'biosecurity' is taken to mean measures aimed at preventing the *deliberate misuse* of the life sciences by non-state actors. In contrast, 'biosafety' is taken to mean measures aimed at preventing the *accidental or unintended misuse* of the life sciences.

suitability of existing regulatory approaches in light of potentially novel risks. However, in the case of recombinant DNA technology, biosafety concerns – notably, concens about the possible unintended consequences of genetically modified organisms (GMOs) – were the primary focus of scientific deliberations at the Asilomar Conference and subsequent policy discussions. In the case of synthetic biology, a field that has emerged at a time of heightened concerns about (bio)terrorism, the possibility that synthetic biology could enable non-state actors to acquire (novel) biological agents that could be used as weapons has been an omnipresent source of concern. In 2009, synthetic biology came to the attention of the US Federal Bureau of Investigation (FBI) and in 2016 gene editing was listed as a potential weapon of mass destruction (WMD) by the US Intelligence Community (Ledford 2010; Clapper 2016).

To more fully understand the top-down governance challenges introduced by synthetic biology it is necessary to consider how synthetic biology's novelties could disrupt or potentially undermine existing biosecurity regulations. In this chapter, we attempt to advance this discussion in several ways. First, we consider the scope and content of existing biosecurity regulations at the international and national levels. Second, we discuss several aspects of synthetic biology that present distinct regulatory challenges. Finally, we conclude with recommendations for strengthening current approaches to top-down biosecurity governance.

Taken together, we argue that although synthetic biology appears to be broadly (if indirectly) covered by existing international and national regulatory systems, several novelties underline the limitations of top-down governance approaches premised upon prohibiting access to specific 'intrinsically dangerous' scientific artifacts (McLeish and Nightingale 2007). Indeed, by some accounts, such restrictions may not only be ineffective, but may also make the world less safe. In an era of synthetic biology – characterized by technology convergence, increased access to bioengineering capabilities, and rapid growth in intangible life science knowledge – top-down governance must be increasingly adaptive, and hybrid forms of governance (incorporating a 'mix' of top-down and bottom-up approaches that leverage the self-governance potential of non-governmental actors) should be encouraged.

3.2 Understanding the Scope and Limitations of Top-Down Governance for Biosecurity and Synthetic Biology

To understand how synthetic biology could challenge or undermine existing approaches to biosecurity oversight and regulation, it is necessary to first consider the international legal instruments relevant to biosecurity, and how these instruments are implemented at the national level (Trump et al. 2020). Based on this analysis, it is apparent that there is no single international legal instrument and no 'one-size-fits-all' approach to national implementation. Rather, the regulatory space governing biosecurity and synthetic biology can most accurately be described as a 'patchwork' of regulatory measures that tend to address biosecurity and synthetic biology indirectly. As scholars have previously observed, the regulatory space governing biosecurity comprises a "collection of cooperative and coercive national and international control measures – including international agreements, multinational organisations, national and international laws, regulations, policies, norms and rules – intended to prevent the spread of dangerous weapons and technologies" (McLeish and Nightingale 2007, p. 1638).

3.2.1 International Instruments for Biosecurity

3.2.1.1 Biological Weapons Convention (BWC)

International legal instruments establish global norms and obligations that are implemented by countries according to their unique risk and regulatory cultures. In relation to biosecurity, no single instrument is more important or directly relevant than the Convention on the Prohibition of the Development, Production, and Stockpiling of Bacteriological (Biological) and Toxin Weapons and on their Destruction, commonly known as the Biological Weapons Convention (BWC). The BWC, which opened for signature on 10 April 1972 and entered into force on 26 March 1975, is the first multilateral disarmament treaty banning the development, production and stockpiling of an entire category of WMD.[3] Under Article I of the Convention, member states agree that they must not "develop, produce, stockpile or otherwise acquire or retain: (1) Microbial or other biological agents, or toxins whatever their origin or method of production, of types and in quantities that have no justification for prophylactic, protective or other peaceful purposes; (2) Weapons, equipment or means of delivery designed to use such agents or toxins for hostile purposes or in armed conflict."

Although tracing its origins to the Cold War, therefore focused on the activities of states and the possibility of biological warfare, the BWC remains relevant and has proven remarkably adaptive in the face of emerging concerns about non-state actors and advances in science and technology (S&T), including synthetic biology. In relation to non-state actors, the BWC requires, as defined under Article IV, States Parties to take any necessary measures "to prohibit and prevent the development, production, stockpiling, acquisition or retention of the agents, toxins, weapons, equipment and means of delivery specified in Article I of the Convention, within the territory of such state, under its jurisdiction or under its control anywhere." In other words, States Parties have a responsibility to enforce the Convention irrespective of who (state or non-state) is acting in contravention to the universal ban on biological weapons.

[3] See: https://www.un.org/disarmament/wmd/bio/

In relation to advances in S&T, the BWC is widely recognized as embodying a 'General Purpose Criterion', whereby the hostile use of biology – irrespective of the specific agents, knowledge or technologies involved – is universally prohibited. This means that new discoveries enabled by advances in S&T, including possible future 'novel' agents produced using synthetic biology techniques or technologies, are covered (Hart and Trapp 2012). In other words, the BWC is effectively 'future proofed' – it "cannot be innovated around, and it embodies the norm in a timeless form" (McLeish and Nightingale 2007, p. 1638).

With a view to advances in synthetic biology, States Parties to the BWC recognize that the field, among other areas of S&T, is rapidly evolving and could potentially introduce novel risks over time (Hart and Trapp 2012). It is equally recognized that developments in S&T could offer new opportunities for countering bioterrorism, and detecting and responding to attacks should they occur (ibid.). To keep pace with advances in S&T and their implications for the BWC, there is general support among States Parties for increased scientific and technical review within the BWC process (ibid.). Notably, proposals have been made for establishing a scientific advisory body that could play a critical role in assessing the impact of advances in S&T on the BWC regime, as well as building consensus among States Parties based on a systematic review of developments in the life sciences relevant to the Convention (ibid.).

While the BWC establishes a global norm against the hostile use of biology, it is nonetheless confronted by a number of challenges and limitations. Notably, despite having 183 States Parties, four Signatory States have yet to ratify the Convention, and ten states have neither signed nor ratified it. There is therefore a need to continue to strive for universality to ensure that the BWC is universally ascribed to and ultimately implemented and enforced at the national level. Moreover, the BWC lacks a verification mechanism to monitor compliance with the Convention. In the absence of such a mechanism, confidence-building measures (CBMs) – voluntary annual reports describing a member state's activities relevant to the Convention – are intended to build trust and transparency. However, annual CBM submissions remain low, the quality of submissions is inconsistent, and States Parties are not obliged to explicitly report on S&T developments (Lentzos and Hamilton 2010). Finally, the BWC is limited by resource constraints and currently depends upon a three-person team – the BWC Implementation Support Unit (ISU) – to facilitate meetings and support daily administrative operations (Hart and Trapp 2012).

3.2.1.2 Chemical Weapons Convention (CWC)

Complementing the BWC, the Convention on the Prohibition of the Development, Production, Stockpiling, and Use of Chemical Weapons and on their Destruction, commonly known as the Chemical Weapons Convention (CWC), prohibits the development, acquisition, possession, transfer and use of toxic chemicals and their precursors for weapons purposes. Unlike the BWC, the CWC, which opened for signature in 1993 and entered into force four years later, is administered by an autonomous international organization, the Organization for the Prohibition of Chemical Weapons (OPCW), which performs a variety of administrative, legal and field functions, including verification to ensure compliance.

Because the BWC and CWC both cover toxins, there exists an overlap between the two conventions and the possibility of mutually reinforcing legal coverage (Hart and Trapp 2012). However, because toxins can be interpreted as biological *or* chemical weapons, situations could arise where states decline to take specific measures to prevent the misuse of toxins under either agreement (ibid.). Advances in synthetic biology – an interdisciplinary field encompassing biology, chemistry, engineering and computing – are likely to make distinctions between chemical and biological weapons even more complex. In this environment, there is a need for inter-convention dialogue to better understand the risks, as well as the jurisdictions and responsibilities of all relevant international conventions and legal instruments.

3.2.1.3 Australia Group, United Nations Security Council Resolution 1540 and Others

Other important international instruments relevant to biosecurity include the Australia Group (AG) and United Nations Security Council Resolution 1540 (UNSCR 1540). In the case of the AG, established in 1985 in response to evidence that Iraq had sourced precursor chemicals and materials for its chemical warfare program through legitimate channels,[4] member states have harmonized export controls covering materials and technologies likely to contribute to the development of chemical or biological weapons. Biological agents and dual-use biotechnology were specifically added to the AG guidelines in 1992 (Oye 2012). In 2008, in light of advances in synthetic biology, the AG established a dedicated advisory body to keep pace with developments in the field and to suggest responses to synthetic biology innovations (ibid.). A key challenge faced by the AG is the growing relevance of intangible technology transfers, which not only make-up an increasingly significant component of legitimate life science research, but also present distinct challenges to regulatory control. Unlike physical pathogens and dual-use equipment, intangible life science transfers cannot be easily monitored and prevented from crossing borders. In the case of synthetic biology, a field characterized as much by digital and informational resources (e.g. DNA sequence information) as physical ones (e.g. DNA sequencers), the regulatory challenges posed by intangible technology transfers are especially acute.

Established in 2004, UNSCR 1540 "obliges States, *inter alia*, to refrain from supporting by any means non-State actors from developing, acquiring, manufacturing, possessing, transporting, transferring or using nuclear, chemical or biological

[4] See 'the origins of the Australia Group'. The Australia Group website. URL [https://www.dfat.gov.au/publications/minisite/theaustraliagroupnet/site/en/origins.html] (accessed 25 March 2021)

weapons and their means of delivery."[5] Under this resolution, member states have specifically agreed to enact relevant legislation and to demonstrate national implementation through national reporting. In addition to directly contributing to global efforts to combat WMD, UNSCR 1540 is significant due to its explicit focus on non-state actors. This focus, as others have observed, "marks a new development in biosecurity policy, which historically has been state-centric" (McLeish and Nightingale 2007, p. 1640). This development not only reflects growing concerns about bioterrorism, but also the belief that advances in S&T have contributed to lowering technical barriers and enhancing the capabilities of non-state actors (ibid.).

In addition to the international instruments discussed above, further international regulations and conventions, including the World Health Organization (WHO) International Health Regulations (2005) (IHR) and the Convention on Biological Diversity (CBD), cover aspects of biosecurity and synthetic biology. In the case of the IHR, the scope of the regulations cover natural, accidental and deliberate disease events, thus capturing biosafety and biosecurity. The WHO, which directs and coordinates international health within the UN system, also monitors and offers guidance on life science research recognized as dual-use research of concern (DURC), including notable experiments involving synthetic biology. In 2012, in response to two such experiments (one led by a team in the Netherlands, the other by a team in the US) that resulted in laboratory-modified H5N1 viruses capable of airborne transmission between mammals ('gain-of-function' experiments), the WHO convened a technical advisory group that considered the biosafety and biosecurity implications of the research, including concerns about the public dissemination of the findings (WHO 2012). In 2015, the WHO convened another scientific working group to address the public health implications of synthetic biology as it relates to smallpox preparedness and control (WHO 2015). The working group concluded that, in light of advances in synthetic biology, including *de novo* DNA synthesis, the risk of smallpox re-emerging can never be fully eradicated. Among the working group's recommendations was the need for revised regulations for research on DNA fragments and the synthesis of virus DNA by new technological approaches.

In the case of the CBD, many considerations that apply to GMOs remain relevant in the case of synthetic biology. Since 2010, the CBD has considered whether synthetic biology should be classified as a new field presenting novel risks and whether new regulations are needed in view of the protection of biodiversity and genetic resources (Lai et al. 2019). Although these deliberations have been oriented to biosafety, the protection of biodiversity and the management of digital sequence information are also relevant for biosecurity.

Taken together, these international conventions and agreements create overlapping governance structures that cover biosecurity and synthetic biology more or less directly. Over time, each of these conventions have evolved and adapted to address emerging risks and regulatory challenges introduced by advances in S&T and non-state actors. As they will undoubtedly continue to evolve and adapt, it will be

[5] See: https://www.un.org/en/sc/1540/about-1540-committee/general-information.shtml

important for all parties to monitor developments in synthetic biology, develop and share common definitions, and determine to what extent synthetic biology introduces novel biosafety and biosecurity risks.

3.2.2 *National Implementation*

International conventions are intent-based, broadly defined and therefore generally future proofed, but they only take effect when they are ratified and implemented at the national level. For example, to implement the BWC, countries must adopt appropriate penal measures criminalizing the production, handling and use of biological weapons;[6] biosafety and biosecurity measures accounting for the safe and secure handling of dangerous pathogens; and import and export controls covering specific biological agents and dual-use equipment and technology.[7] Finally, enforcement measures must be adopted to ensure the ongoing monitoring of life science activity and to prosecute and punish offenders.[8] How countries carry out each of these implementation measures is influenced by a variety of factors, including a country's attitudes toward risk and the importance of technological innovation. In practice, countries often draw on and/or adapt existing laws and regulations, rather than creating new ones. For example, aspects of the BWC may already be covered by existing criminal laws, public health (and medical) laws, emergency management laws and/or national security laws (Fidler 2001; Colussi 2015). This means that national implementation not only takes different forms *between* countries, but also tends to result in a patchwork of rules applying directly, or more often indirectly, to biosecurity and synthetic biology *within* countries.

With a view to national implementation, this section discusses some general characteristics of top-down approaches to biosecurity and synthetic biology regulation. There are numerous categorizations that can be used to group these approaches, but for the purpose of this chapter we limit ourselves to the distinction between hard and soft law, between different levels of precaution in governance, and between general and specific legislation. It should be emphasized that these distinctions are merely ideal types. While some examples of national implementation will be given, it should be noted that most countries apply a mix of different strategies that cannot be strictly assigned to a single type of governance.

---

[6] See for example The Biological Weapons Act (1974) in The UK or the Biological Weapons Act (1989) in the USA.

[7] See VERTIC factsheet on national implementation measures for the BWC, available at: http://www.vertic.org/wp-content/uploads/2019/08/FS7_BWC_EN_FEB_2018.pdf

[8] Ibid.

3.2.2.1 Hard Law vs. Soft Law

When selecting regulatory options, national governments have to balance *hard law*, meaning statutorily defined legal prescriptions that result in punishment when violated, and *soft law*, comprised of norms and standards within the operating community that build validity, trust, and collaboration opportunities between community members and other stakeholders. By these definitions, hard law is generally taken to correlate with traditional 'top-down' governance. However, a government can equally make the deliberate 'top-down' choice to cede some responsibility to soft law in an effort to stimulate ownership and self-responsibility on the part of industry and technology users, encourage economic development, limit the burden and costs of regulatory oversight, etc. (Palumbo and Bellamy 2010). The balancing of hard and soft law can be categorized according to three generalized types of governance approach: precautionary, laissez-faire and stewardship (Linkov et al. 2018a, b).

Precautionary Governance

There is a broad range of precautionary approaches that are discussed in the scholarly literature and that have been implemented by governments in practice (Ahteensuu and Sandin 2012; Dinneen 2013). Rather than analyzing each of these approaches in detail, the following discussion considers precautionary governance more generally, treating it as a set of pre-emptive regulations aimed at ensuring the safe and secure application of technologies and preventing exposure to risk (Linkov et al. 2018a, b).

In general, precautionary governance is associated with risk-aversion and centralized governance systems that require safety to be demonstrated prior to permitting the use of new technologies or products (Stirling 2006). For example, the European Union (EU) is widely known to take a precautionary approach in relation to GMO biosafety as prescribed by EU legislation (European Commission 2000; Anyshchenko 2019). This legislation, which aims to protect the health and safety of humans, animals, and the environment from adverse biological contamination, equally applies to aspects of synthetic biology (especially genome editing).[9]

Comparable legal requirements for biosecurity were introduced later in Europe (2000),[10] and much of the GMO debate remains focused on the subject of biosafety. Following a recent review process on whether or not existing gene technology regulations and risk assessment and management practices are applicable to synthetic biology, three opinion statements by the Scientific Committees did not address

[9] See Case C-528/16, Court of Justice of the European Union 2018. Organisms obtained by mutagenesis are GMOs and are, in principle, subject to the obligations laid down by the GMO Directive. However, whereas radiation and chemically induced mutagenesis are exempt from the regulations because they have a so called 'history of safe use', mutagenesis induced by gene editing techniques such as CRISPR is not because it lacks this 'history of safe use'.

[10] See Regulation 1334/2000 of 22nd June 2000 setting up a Community regime for the control of exports of dual-use items and technology in O.J. L 159/2000, modified by Regulation 2432/2001 of 20th November 2001 in O.J. L 338/2001, and by Regulation 428/2009 of 5th May 2009 in O.J. L 134/2009.

biosecurity directly, focusing solely on biosafety (SCHER, SCENIHR and SCCS 2014; SCENIHR, SCHER and SCCS 2015). By contrast, synthetic biology's biosecurity implications [11] have been the subject of considerably more policy discussion in the US (e.g. NSABB 2006, 2011; for a review, see Oye 2012). Whereas Europe has historically strongly focused on biosafety, biosecurity regulations were developed in the US at a relatively early stage (1989). Moreover, since 9/11 and the subsequent anthrax letter attacks, US policy has further emphasized biosecurity (McCarty 2018). These developments have resulted in extensive controls on scientific research (McLeish and Nightingale 2007).[12]

A precautionary governance system seeks to protect against undue and unnecessary harm, but this approach can also impose costs. It is often argued that a strict pre-emptive regulatory system can potentially cut off avenues for innovation and industry and diminish a country's economic development and international competitiveness. These potential pitfalls have been pointed out both in relation to GMO biosafety regulations in Europe and biosecurity controls in the US (Wager and McHughen 2010; Bogner and Torgersen 2018; Gaudioso and Salerno 2004). Moreover, in a world where other countries may be more risk-tolerant, countries that adopt a precautionary governance approach are not necessarily insulated from risk. This problem has been faced before when individual countries have pursued, for example, nuclear disarmament or sought to reduce or eliminate nuclear power plants domestically while neighbors have not. Thus, strict adherence to precautionary governance in the case of biosecurity and synthetic biology is not a panacea for all threats.

Laissez-Faire Governance

A laissez-faire governance approach cedes much of the regulatory power to existing or emerging bottom-up initiatives, placing trust in the capacity of technology producers, industry and users to play an active role in their own regulation. Under this approach, such non-governmental actors are encouraged to determine (at least in part) how safety and security practices are structured, implemented and enforced, while centralized government plays a role in setting minimum standards and intervening in the event of regulatory failures. This approach is generally intended to promote innovation and flexibility, as well as rapid adaption and response to emerging threats (Linkov et al. 2018a, b).

While laissez-faire governance is a promising approach that recognizes the important role that non-governmental actors can play in the regulatory process, there are also potential pitfalls. One such example, albeit focused on bioethics rather than biosecurity, can be traced to the use of germline editing in humans to produce the first CRISPR baby. In this case, despite broad international agreement that scientists should *"hold off on editing human eggs, sperm or embryos until gene-editing*

---

[11] See Ahteensuu 2017.

[12] For a review of all the legislative framework about bioterrorism in the U.S.A., see RICHARDS, Edward P./O'Brien, Terry/Ratburn, Katharine C., "Bioterrorism and the Use of Fear in Public Health", The Urban Lawyer, No. 3, Vol. 34, 2002, pp. 685–726.

technology (and the implications of the edits) are better understood."[13] (See also Cyranoski and Ledford 2018), a researcher in China took advantage of laissez-faire regulatory controls that resulted in multiple potential risks and unintended consequences. These included the possibility of long-term changes to the human germline; encouraging other scientists (including those working internationally) to pursue germline editing in humans (Cyranoski 2019a, b), and motivating Chinese regulators to introduce stricter regulatory controls on genetic research (ibid.).

As the above example suggests, one risk of laissez-faire governance is that an individual's risk tolerance may not (intentionally or unintentionally) conform to existing norms and their actions may subsequently expose everyone to undue risk or irreversible harm, with implications for the laissez-faire state as well as other states. Additionally, the more a state relies on soft law, the more responsibility the government delegates to individuals, groups and organizations, not only to establish norms and follow them, but also to enforce them. In the case of synthetic biology, such bottom-up initiatives have played an important role in this rapidly developing field. Organizations such as the International Gene Synthesis Consortium (IGSC) have come to play an important role in the regulatory process by, for example, developing industry standards and guidelines (IGSC 2017). For better or worse, the success of such approaches will depend on the commitment of non-governmental actors to act in the best interests of society, valuing safety and security as a public good.

Ultimately, while a laissez-faire approach may effectively supplement aspects of centralized government regulation, it cannot be expected to fully replace it. This is because, as history has shown, individual researchers or individual members of industry will sometimes choose to value personal prestige or cost cutting over safety and security. Moreover, from the standpoint of non-governmental actors, including the DNA synthesis consortia noted above, regulation is not necessarily a bad thing. Indeed, by some accounts, the standards and codes of conduct produced by the DNA synthesis industry were motivated by a lack of top-down regulations that could provide a benchmark for not only mitigating potential biosecurity risks, but also liability issues and reputational costs in the event of an incident. For this reason, the US Government's own DNA screening guidance[14] has been largely welcomed by industry.

Stewardship Governance

A stewardship governance approach seeks to balance the advantages of laissez-faire governance with centralized risk management, and different countries may enact stewardship approaches that incline towards one side or the other (Linkov et al. 2018a, b). Governments that adopt a stewardship governance approach to synthetic

---

[13] Many articles and newspaper items have in response to the Jiankui He case referred to the (perceived) broad worldwide consensus that germline editing in the clinic would be a step too far. See a.o. Weintraub (2019).

[14] See: https://www.phe.gov/Preparedness/legal/guidance/syndna/Documents/syndna-guidance.pdf

biology will seek to monitor developments in the field, enabling space for innovators and industry to operate, while intervening in the event of observed inconsistencies with objectives formulated through multi-stakeholder processes involving both government and non-governmental actors. Stewardship governance is more guidance than direction, and typically involves the active participation of bottom-up entities in formulating norms alongside traditional top-down regulatory bodies. Though some hard constraints exist, they are likely to arise through consensus driven collaborative forums that both support beneficial innovation and use of technologies while critically evaluating risks (the Goldilocks principle).

Many countries apply a stewardship style of governance for emerging technologies, including synthetic biology. While the specific approaches of individual countries differ at the level of detail, they are all based on the principles of being flexible and adaptive and using a mix of different tools to achieve a 'balanced' form of governance. For example, the US employs a stewardship model in much of its engagement with emerging technology developers and users: the executive government, state governments, the US patent system, insurers, and the system of legal liability all serve to foster innovation while constraining what scientists can attempt within established risk tolerances. The stewardship model aims to limit innovation only when the risks are deemed sufficient to justify government intervention.

Similar governance concepts, capturing the notion of balancing laissez-faire and precautionary models, mentioned in the literature include 'adaptive governance' and 'prudent vigilance'. The main characteristic of adaptive governance is its emphasis on flexibility, allowing "stakeholders in industry, government, and society at large to iteratively adjust their best practices and codes of conduct to derive the benefits…without incurring unnecessary or unacceptable risks or losses" (Linkov et al. 2018a, b). The 'prudent vigilance' model "establish[es] processes for assessing likely benefits along with safety and security risks both before and after projects are undertaken" (PCBSI 2010). Introduced by the US Presidential Commission for the Study of Bioethical Issues in its policy report on synthetic biology, the model reflects a combination of 'top-down' and 'bottom-up' strategies for the enforcement and control of biosecurity risks associated with synthetic biology (see Collussi 2015).

3.2.2.2 Biosafety vs. Biosecurity, GMOs vs. Synthetic Biology

Given that biosecurity and synthetic biology are relatively new concepts with limited regulatory legacies, there are few sign posts to indicate how countries' regulatory frameworks should evolve to meet their associated risks, and even if new regulations are needed. In light of existing GMO and biosafety regulations, countries are faced with the question of determining what is already covered and what is not. In this context, much depends on the 'newness' or 'novelty' ascribed to synthetic biology relative to conventional biotechnology, and governments must weigh the benefits and costs of introducing new regulations that are typically complex and time-consuming to produce (Hamilton 2015).

Even in states where synthetic biology is actively pursued, many do not (yet) explicitly reference it in legal documentation. In Germany, for example, the Research Office of Parliament concluded in 2015 that the processes currently called 'synthetic biology' are in fact still conventional biotechnology and can be covered by existing regulations, including existing security, transport, and export control regulations, and the Health Care Act, for issues related to human health. From the standpoint of the Research Office of Parliament, synthetic biology results in new genetic combinations of a host organism with a variable amount of new genetic material, just like GMOs. Similar conclusions have been reached by scientific advisory bodies in Europe and beyond (Trump 2017; ZKBS 2018; Pauwels et al. 2013).

Synthetic biology is similarly not mentioned in EU legislation and, historically, changes to the EU biosafety regulations have proven to be extremely time consuming due to different perspectives on both the necessity and desirability of change (Eriksson et al. 2018). In the case of several new plant breeding techniques, it remains unclear whether they are covered by EU GMO legislation, despite deliberations (including several scientific and legal expert committees and reports) stretching back more than 15 years.

Finally, distinctions between biosecurity and biosafety can be ambiguous. In countries with regulations specifically covering biosecurity, a variety of definitions can be found depending on the context and field of application. For example, different interpretations exist between the human health sector, and the animal and plant health sectors . In the case of human health, biosecurity is generally understood as a set of regulatory measures aimed at preventing the deliberate misuse of biology (i.e. the same way we interpret biosecurity in this chapter). In the case of animal and plant health, biosecurity is generally understood as a set of regulatory measures aimed at preventing and responding to the natural or unintentional introduction, establishment and spread of pests or pathogens (Mumford et al. 2017).[15] Some languages also lack distinct words for biosafety and biosecurity (at least in general usage). For example, *biosecuridad,* in Spanish, *Biosicherheit,* in German, and *bioturvallisuus,* in Finnish, are generally used to capture both biosafety and biosecurity. This can result in misunderstandings, as these concepts can be understood differently between countries and between sectors. In the case of languages that do differentiate between biosafety and biosecurity, there can nonetheless be confusion

[15] In the setting of the BWC, it is most commonly used to refer to mechanisms to establish and maintain the security and oversight of pathogenic microorganisms, toxins and relevant resources. But For example, the glossary of the FAO Basic Laboratory Manual for the Small-Scale Production and Testing of I-2 Newcastle Disease Vaccine considers biosecurity to be "precautions taken to minimize the risk of introducing an infectious agent into a population". And in the glossary of the New Zealand Parliamentary Commissioner for the Environment considers biosecurity to be "The exclusion, eradication and effective management of pests and unwanted organisms into New Zealand." http://www.pce.govt.nz/reports/pce_reports_glossary.shtml. The OECD developed best practice guidelines for biosecurity at ancillary facilities, defining it as "institutional and personal security measures and procedures designed to prevent the loss, theft, misuse, diversion or intentional release of pathogens, or parts of them, and toxin-producing organisms, as well as such toxins that are held, transferred and/or supplied by Biological Resource Centres".

over how each set of practices should be applied and what actors or institutions bear responsibility for their oversight. Recognizing these ambiguities, initiatives in the EU and Central Asia have been undertaken to map how these terms are understood and applied in different countries (EBRF 2016; EEAS 2017).

3.2.2.3 National Implementation: An Inevitable Patchwork

In many (if not most) countries, the regulatory frameworks governing synthetic biology and biosecurity are fragmented – divided across multiple pieces of legislation (e.g. legislation on bioweapons, dual-use materials and technologies; export/import and transport; gene technology; human health; microorganisms, animal and plant health; agriculture; occupational health; waste disposal; criminal behavior, etc.) – and tend to address one or both subjects only indirectly (Greer and Trump 2019). In Finland, for example, over twenty acts and regulations can be interpreted as governing the biosafety and biosecurity dimensions of synthetic biology, and none make explicit reference to 'synthetic biology'. In other countries, like the US, synthetic biology is explicitly referenced in relevant regulations and guidance documents (e.g. federal DNA screening guidance), accompanied by more than 35 different biosecurity regulations (some of which have been described as mutually inconsistent, making compliance with all of them impossible): "The regulation of products of synthetic biology is juggled, and not always clearly so, among three federal agencies, various federal laws, and the Coordinated Framework (…). The regulatory framework that has evolved is complicated, increasingly circuitous, and not for the faint of heart" (Bergeson et al. 2015).

In other countries, especially developing countries that may share very different priorities due to limited resources and urgent challenges associated with human rights and food security, relevant biosecurity and synthetic biology regulations have not yet been adopted. Thus, the fact that international conventions are in place may create a false sense of confidence about the level of consensus and adoption. In practice, the effectiveness of conventions depends on how they are implemented, a task that can take a considerable amount of time. Indeed, despite the relatively long history of GMO regulation, relevant laws have not been adopted by all countries.

Additionally, in an interconnected and globalized world, the effectiveness of national implementation is limited in its ability to prevent or limit access to new technologies that may be carelessly used or transferred by other states. National implementation acts within legal and geographical boundaries and depends upon rules designed to shape the behavior of domestic audiences. For truly robust governance to occur, all states must work to mitigate the risks posed by advances in synthetic biology through effective national implementation.

Given the diversity of national implementation efforts globally, it seems unlikely that there will be a unified approach and that all gaps in the governance of synthetic biology will be filled. However, potential vulnerabilities can be addressed through a combination of different instruments. "Often, approaches to risk governance are defined in terms of a choice between two alternatives. Either accept the precautionary principle but in so doing choke off development of potentially promising technologies, or go with laissez-faire and in so doing accept potentially irreversible harms" (Oye 2012, p. 22). Linkov, Trump, Poinsatte-Jones, and Florin (2018b) emphasize the importance of a stepwise learning approach under conditions of acknowledged uncertainty, with initial limits on use, iterative phases of data gathering and regulatory evaluation. In addition to adopting hybrid governance models, combining elements of precaution with policies aimed at stimulating innovation, governments may also look to strengthen regulatory systems through a combination of hard and soft law. For example, legal measures can be complemented by codes of conduct or guidelines produced by researchers and industry.

Top-down governance systems, in their various forms, offer advantages and disadvantages. The regulatory challenges presented by synthetic biology will require the careful consideration of multiple (hybrid) governance options.

3.3 Key Novelties and Tensions Introduced by Synthetic Biology

Biosecurity in the context of synthetic biology benefits from the groundwork previously laid to mitigate biological weapons threats and the risks posed by earlier advances in biotechnology. We have seen that international conventions and national implementation already cover (albeit imperfectly) many aspects of synthetic biology. In this section, we discuss three aspects of synthetic biology that represent distinct governance challenges: convergence, democratization and intangibility. Cutting across these tensions are overarching issues that are familiar to all emerging technology discussions, including the pace of technological change and uncertainty, both in terms of the potential risks and benefits (Marchant et al. 2011). For effective governance in this environment, regulatory efforts must seek to be forward-looking and adaptive. Moreover, whether states are applying existing legislation to synthetic biology or enacting new legislation, both relevance and coherence will be paramount.

3.3.1 Convergence

Because synthetic biology represents a convergence between biology, chemistry, engineering and computing, ambiguities may arise regarding which conventions should regulate specific developments. This could produce redundancies in regulatory efforts or, far worse, gaps in responsibility as each authority presumes another's attention and jurisdiction. Frontier research on protocells and xenobiology, among other domains of synthetic biology that push the limits of scientific classification, may even fall outside the scope or remit of existing conventions and established legislation. Other synthetic biology risks, including those that blur the line between biological and informational hazards, may require new security concepts and practices. For example, recent studies that have demonstrated the capacity to encode computer viruses in synthesized strands of DNA, exploiting vulnerabilities in the sequencing and processing pipeline (Ney et al. 2017), underline the growing relevance of *cyberbiosecurity*, an emerging field at the intersection of cybersecurity and biosecurity.

Synthetic biology processes and methods may also become (or may already *be*) too diverse to legislate individually, and the convergent nature of synthetic biology may lead to fragmentation or duplication of laws at the national level. In this environment, it may become increasingly unclear which laws should be applied to synthetic biology and how judges or legislators should interpret and apply them. In light of its numerous applications, multiple contributing scientific disciplines, and practitioners working in both institutional and non-institutional settings, synthetic biology is an exemplary case of convergence.

3.3.2 Democratization

Synthetic biology provides new ways to modify organisms outside of dedicated laboratories and without advanced skills (Oye 2012). This broadens access to the science and enables individuals to apply its techniques without oversight from formal institutions or associated institutional norms (NSABB 2011; Gruber 2019). Theoretically, democratization could allow untrained or malicious actors to create dangerous organisms. In practice, current synthetic biology applications are far from facilitating this: in most cases, so-called do-it-yourself (DIY) biology is limited to relatively simple experiments with nonpathogenic organisms, with relatively low success rates (Kuiken 2016). However, looking to the future, it is possible that advances in synthetic biology will lower the technical barriers needed to engage in more advanced bioengineering projects. Much like the history of computing, as synthetic biology tools and techniques become more reliable, streamlined, and easy to use, bioengineering may become common place, accessible to specialists and non-specialists alike.

If synthetic biology does (eventually) make biology easy or at least significantly *easier* to engineer, traditional top-down governance and enforcement will no longer be sufficient to provide adequate oversight, and there is likely to be a growing need to enlist the support of the synthetic biology community itself to participate in various forms of self-regulation or self-policing. Already, innovative self-governance approaches of this kind have been employed in the context of DIY-biology, wherein DIY-culture and social protocols have been leveraged to support responsible science and self-regulation (Bolton and Thomas 2014). However, such regulation works only if all parties engage, or are permitted to engage, which is not always guaranteed.

One potential model for policing synthetic biology in an era of democratization is provided by the partnership that has evolved between the International Genetically Engineered Machine (iGEM) competition and the US FBI. This partnership has sought to create a culture of trust and transparency between law enforcement and members of the synthetic biology community through engagement with students and the private sector, demonstrating positive results (Ossola 2016). Moreover, iGEM organizers require, as a condition of student teams' participation, students to engage with safety and security issues throughout their projects (Oye 2012). These partnership activities create an opportunity for mutual learning, but there can be tensions. In 2019, 47 countries sent teams to iGEM, but some international students were barred from attending the event (Baber 2018) due to a US Executive Order that banned students from several countries, including Iran and Syria. This represented a collision of top-down priorities. The travel ban's focus on national security compromised the FBI's ability to build relationships within the evolving international synthetic biology community.

Democratization in synthetic biology requires balancing individual liberty and risk prevention. In a hypothetical future world of broad bioengineering capabilities, safeguards will still be necessary, but whether they are best applied through top-down or bottom-up efforts remains to be seen. The successful governance of synthetic biology will partially depend upon the functioning collaboration between top-down and bottom-up governance in identifying and preventing purposeful or accidental misuse. Bottom-up governance is further discussed in Chap. 6 of this volume.

3.3.3 Intangibility

The 'ingredients' for synthetic biology are increasingly informational, thus regulations focusing on material control, while important, cannot address the full scope of synthetic biology's risks. Digital sequence information, access to online research protocols and methodologies, and the capacity to construct laboratory hardware from scratch using 3D-printing technologies are all developments that threaten to undermine regulatory systems that privilege policies aimed at restricting access to physical technologies (NRC 2004). In an era of the life sciences dominated by the production and distribution of information-based resources, effective regulatory controls on intangible technology transfers are essential. However, successfully designing and implementing controls of this kind is a challenging task. Whereas dual-use equipment must (or at least should) pass through physical checkpoints, digital sequence information can be transmitted with the click of a button.

Advances in synthetic biology may also yield threats that are not only difficult to regulate, but impossible to *anticipate*. For example, it may someday be possible to design and build novel genomes (based on existing, modified or new genetic code) that transcribe previously unknown pathogens. Therefore, although select agent lists are likely to remain relevant (if for no other reason than because they generate awareness about pathogens that *are known* to cause severe harm to public, animal or plant health), they cannot be expected to capture the full spectrum of harmful agents that are (or may someday be) possible to create.

Finally, in the context of synthetic biology and the contemporary life sciences in general, important questions remain about what life science information should be considered 'risky' in the first place and how this information aught to be controlled. To place these considerations in context, one need look no further than the H5N1 gain-of-function experiments (Imai et al. 2012). In this case, concerns were raised about whether research describing the synthesis of a novel H5N1 variant was suitable for open publication. Some argued publishing the protocols would provide a blueprint for bioterrorism. Others asserted that the research should never have been conducted. Decisions about whether (or what parts of) the research should be published were sources of international debate and global controversy (Hamilton 2015). While past technologies have motivated similar controversies (see McLeish and Nightingale 2007), questions about the intrinsic dangers of life science information, and what information may be too dangerous to share, have never been more acute.

3.4 Conclusions and Recommendations

- The regulatory landscape for biosecurity and synthetic biology can best be described as a 'patchwork' of international conventions, national laws, regulations, guidelines, etc. In many instances, these were designed to address other (state biowarfare programs) or earlier (biosafety) concerns.
- While synthetic biology appears to be broadly (if indirectly) covered by existing top-down governance measures (e.g. GMO laws), several characteristics of the science, including convergence, democratization and intangibility, point to possible regulatory gaps. How governments address these novelties depends upon their regulatory cultures and perceptions of risk.
- To date, the regulatory response, while varied (ranging from more precautionary to more laissez-faire), suggests a preference for evolutionary rather that revolutionary regulatory change. Like the regulatory response to GMOs, there is a tendency for governments to adapt existing regulations to new technologies.
- Whether new conventions, laws or regulations are (or are not) needed to address synthetic biology's novelties is open to question. At the very least, there is a need to monitor advances in the field and to consider how top-down governance approaches could be improved. The following recommendations aim to advance this discussion:
- The BWC, the premier international forum that addresses biological threats, should play a leading role in monitoring security-relevant advances in synthetic biology. Proposals to establish a BWC scientific advisory body and to introduce a S&T reporting requirement into the CBM mechanism should be encouraged.
- In view of the growing convergence between biology, chemistry, engineering and computing, inter-convention dialogue is needed between the BWC and CWC, among others, to ensure the full scope of synthetic biology's risks are taken into consideration and that there is agreement on how to address these risks in the event of deliberate misuse by state or non-state actors.

- Given that many security concerns about synthetic biology relate to its informational (e.g. digital sequence information) rather than physical (e.g. DNA sequencers) dimensions, it is necessary to develop improved methods of regulating intangible technology transfers. It is no longer sufficient to rely exclusively on material controls and list-based approaches to regulation.
- Synthetic biology is contributing to the democratization of genetic engineering. It is therefore essential to enlist the support of non-governmental organizations and actors, including technology developers, industry and users, in the regulatory response. The value of complementing top-down governance measures with bottom-up governance measures, drawing on limited forms of self-regulation or self-policing, will only increase as the tools of modern biology become more accessible.
- There is no one-size-fits-all approach to synthetic biology's governance. Finding the appropriate 'mix' of top-down and bottom-up regulatory measures will require foresight, broad dialogue, and a willingness on the part of governments to look to new, hybrid forms of risk regulation.

Acknowledgements The authors would like to kindly acknowledge the contributions of the following individuals: **Thomas Binz**, Bundesamt für Gesundheit BAG, **Janet Martha Blatny**, Norwegian Defence Research Establishment; **Peter Emanuel,** US Army Futures Command; **Daniel Feakes**, Biological Weapons Convention Implementation Support Unit, United Nations Office for Disarmament Affairs; **Ricardo Gent**, German Association of Biotechnology Industries; **Christian Haggenmiller**, German Institute for Defence and Strategic Studies; **Manuel Porcar**, University of Valencia; **Alessandra Salamini**, Bayer Crop Science; **Alla I. Yemets**, Institute of Food Biotechnology and Genomics, National Academy of Sciences of Ukraine, and **M. Carmen de Vicente,** European Commission. Their valuable contributions as members of the "Top-Down Governance Working Group", within the framework of the "NATO Science for Peace and Security (SPS), Advanced Research Workshop (ARW) on Security for Emerging Synthetic Biology Threats" workshop, 7-11 July 2019, helped frame and inform this chapter.

References

Ahteensuu M (2017) Synthetic biology, genome editing, and the risk of bioterrorism. Sci Eng Ethics 23(6):1541–1561

Ahteensuu M, Sandin P (2012) The precautionary principle. In: Hillerbrand R, Sandin P, Roeser S, Peterson M (eds) Handbook of risk theory: epistemology, decision theory, ethics and social implications of risk. Springer Netherlands, Dordrecht, pp 961–978

Anyshchenko A (2019) The precautionary principle in EU regulation of GMOs: socio-economic considerations and ethical implications of biotechnology. J Agric Environ Ethics 32(5–6):855–872

Baber S (2018) Visa issues prevent several international students from attending iGEM Giant Jamboree. The Tech, 8 November 2018. Available: https://thetech.com/2018/11/08/igem-competition-students. Last accessed 26 Sept 2019

Bergeson LL et al (2015) The DNA of the U.S. Regulatory system: are we getting it right for synthetic biology. Woodrow Wilson Center, October 2015

Bogner A, Torgersen H (2018) Precaution, responsible innovation and beyond–in search of a sustainable agricultural biotechnology policy. Front Plant Sci 9:1884

Bolton R, Thomas R (2014) Biohackers: the science, politics, and economics of synthetic biology. Innov Technol Govern Global 9(1–2): 213–219

Cello J, Paul AV, Wimmer E (2002) Chemical synthesis of poliovirus cDNA: generation of infectious virus in the absence of natural template. Science 297(5583):1016–1018

Clapper JR (2016) Worldwide threat assessment of the US Intelligence Community. Statement for the Record. Senate Armed Services Committee, 9 February 2016. https://www.dni.gov/files/documents/SASC_Unclassified_2016_ATA_SFR_FINAL.pdf

Colussi IA (2015) Synthetic biology as a new threat to biosecurity. Is there a road to suitable governance? Bioterrorismo y bioseguridad 65–110

Cyranoski D (2019a) China set to introduce gene-editing regulation following CRISPR-baby furore. Nature, 21 May 2019

Cyranoski D (2019b) Russian 'CRISPR-baby' scientist has started editing genes in human eggs with goal of altering deaf gene. Nature News, 18 October 2019

Cyranoski D, Ledford H (2018) Genome-edited baby claim provokes international outcry. Nature 563(7733):607–608

Dinneen N (2013) Precautionary discourse. Politics Life Sci 32(1):2–21

Endy D (2005) Foundations for engineering biology. Nature 438(24):449–453

Eriksson D et al (2018) A welcome proposal to amend the GMO legislation of the EU. Sci Soc 36(11):1100–1103. https://doi.org/10.1016/j.tibtech.2018.05.001

European Biosecurity Regulators Forum (2016) Securing immaterial technology with dual-use potential (working paper). http://www.ebrf.eu/documents.html

European Commission (2000) Communication from the Commission on the precautionary principle/*COM/2000/0001 final*/, 8 See also: https://eur-lex.europa.eu/legal-content/EN/ALL/?uri=CELEX%3A52000DC0001

European External Action Service (2017) EU funded project on biosafety and biosecurity kicks-Off regionally in Central Asia. Press release 02/02/2017. UNIQUE ID: 170202_11

Fidler DP (2001) The globalization of public health: the first 100 years of international health diplomacy. Bullet World Health Organ 79:842–849

Garfinkel MS, Endy D, Epstein GL, Friedman RM (2007) Synthetic genomics: options for governance. Available at: http://www.jcvi.org/

Gaudioso J, Salerno RM (2004) Biosecurity and research: minimizing adverse impacts. Science 304(5671):687–687

Greer SL, Trump B (2019) Regulation and regime: the comparative politics of adaptive regulation in synthetic biology. Policy Sci 52(4):505–524

Gruber K (2019) Biohackers: a growing number of amateurs join the do-it-yourself molecular biology movement outside academic laboratories. EMBO Rep 20(6):e48397

Hamilton RA (2015) Governing through risk: synthetic biology and the risk management process. PhD thesis, The London School of Economics and Political Science (LSE)

Hart J, Trapp R (2012) Science, technology, and the biological weapons convention. Arms Control Today 42(8):15–21

Imai M et al (2012) Experimental adaptation of an influenza H5 HA confers respiratory droplet transmission to a reassortant H5 HA/H1N1 virus in ferrets. Nature 486(7403):420–428

International Gene Synthesis Consortium (2017) Harmonized screening protocol V2.0. Gene Sequence & Customer Screening to Promote Biosecurity, 19 November 2017

Jefferson C, Lentzos F, Marris C (2014) Synthetic biology and biosecurity: challenging the "myths". Front Public Health 2:115

Kuiken T (2016) Should we fear DIY biologists' use of cutting-edge gene-editing technology? Scientific American, 18 March 2016

Lai H et al (2019) Synthetic biology and the United Nations. Sci Soc 37(11):1146–1151. https://doi.org/10.1016/j.tibtech.2019.05.011

Ledford H (2010) Garage biotech: life Hackers. Nature 467:650–652. https://doi.org/10.1038/467650a

Lentzos F, Hamilton RA (2010) *BWC Confidence Building Measures: Preparing for a comprehensive review of the CBM mechanism at the Seventh BWC Review Conference*. London School of Economics

Linkov I, Trump BD, Anklam E, Berube D, Boisseasu P, Cummings C et al (2018a) Comparative, collaborative, and integrative risk governance for emerging technologies. Environ Syst Decis 38:170–176

Linkov I, Trump BD, Poinsatte-Jones K, Florin MV (2018b) Governance strategies for a sustainable digital world. Sustainability 10(2):440. Decisions 38(2):170–176

Marchant GE, Allenby BR, Herkerts JR (2011) The growing gap between emerging technologies and legal-ethical oversight: the pacing problem. Springer, New York

McCarty A (2018) Changes in U.S. biosecurity following the 2001 Anthrax attacks. J Bioterror Biodef 9(2). https://doi.org/10.4172/2157-2526.1000163

McLeish C, Nightingale P (2007) Biosecurity, bioterrorism and the governance of science: the increasing convergence of science and security policy. Res Policy 36:1635–1654

Mumford JD, Leach AW, Holt J, Suffert F, Sache I, Moignot B, Hamilton RA (2017) Integrating crop bioterrorism hazards into Pest risk assessment tools. In: Gullino ML, Stack JP, Fletcher J, John D (eds) Practical tools for plant and food biosecurity. Springer, Mumford

National Research Council (2004) Biotechnology research in an age of terrorism. The National Academies Press, Washington, DC. https://doi.org/10.17226/10827

National Science Advisory Board for Biosecurity (2006) Addressing biosecurity concerns related to the synthesis of Select Agents. Report of the National Science Advisory Board for Biosecurity (NSABB), December 2006

National Science Advisory Board for Biosecurity (NSABB) (2011) Strategies to educate amateur biologists and scientists in non-life science disciplines about dual use research in the life sciences. Report of the National Science Advisory Board for Biosecurity, June 2011

Ney P, Koscher K, Organick L, Ceze L, Kohno T (2017) Computer security, privacy, and {DNA} sequencing: compromising computers with synthesized {DNA}, privacy leaks, and more. In 26th {USENIX} Security Symposium ({USENIX} Security 17): 765–779

Noyce RS, Lederman S, Evans DH (2018) Construction of an infectious horsepox virus vaccine from chemically synthesized DNA fragments. PLoS One 13(1):e0188453

Ossola A (2016) Why is the FBI reaching out to student bioengineers? Vice, 26 November 2016. Available: https://www.vice.com/en_us/article/yp3knm/igem-fbi. Last accessed 26 Sept 2019

Oye K (2012) Proactive and adaptive governance of emerging risks: the case of DNA synthesis and biotechnology. IRGC, Public Sector Governance of Emerging Risks, Synthetic Biology Case, June 2012

Palumbo A, Bellamy R (2010) From government to governance. Taylor & Francis LTD, New York

Pauwels et al (2013) Event report: SynBio workshop (Paris 2012) – risk assessment challenges of synthetic biology. J Verbr Lebensm 8(3):215–226

Presidential Commission for the Study of Bioethical Issues (2010) New directions. The ethics of synthetic biology and emerging technologies, December 2010

Regalado A (2016) Top U.S. intelligence official calls gene editing a WMD threat. MIT Technology Review, 9 February 2016. https://www.technologyreview.com/s/600774/top-us-intelligence-official-calls-gene-editing-a-wmd-threat/

SCENIHR, SCHER and SCCS (2015) Final opinion on synthetic biology III: risks to the environment and biodiversity related to synthetic biology and research priorities in the field of synthetic biology. European Union 2015. https://doi.org/10.2875/590512

SCHER, SCENIHR and SCCS (2014) Opinion on synthetic biology 1: definition. European Union 2014. https://doi.org/10.2772/76553

Stirling A (2006) Precaution, foresight and sustainability: reflection and reflexivity in the governance of science and technology. Reflexive Gov Sustain Dev 225

Trump BD (2017) Synthetic biology regulation and governance: lessons from TAPIC for the United States, European Union, and Singapore. Health Policy 121(11):1139–1146

Trump BD, Galaitsi SE, Appleton E, Bleijs DA, Florin MV, Gollihar JD et al (2020) Building biosecurity for synthetic biology. Mol Syst Biol 16(7):e9723

Wager R, McHughen A (2010) Zero sense in European approach to GM: the European Union's zero-tolerance of trace amounts of unapproved genetically modified material in imported food and feed is scientifically unsound and could lead to economic ruin. EMBO Rep 11(4):258–262

Weintraub K (2019) Scientists call for a moratorium on editing inherited genes. Scientific American, Policy and Ethics. https://www.scientificamerican.com/article/scientists-call-for-a-moratorium-on-editing-inherited-genes/

World Health Organization (2012) Report on technical consultation on H5N1 research issues. Geneva, 16–17 February 2012. Available at: https://www.who.int/influenza/human_animal_interface/mtg_report_h5n1.pdf. Accessed 18 Jan 2020

World Health Organization (2015) The independent advisory group on public health implications of synthetic biology technology related to smallpox. WHO/HSE/PED/2015. World Health Organization, Geneva. Available at: https://apps.who.int/iris/bitstream/handle/10665/198357/WHO_HSE_PED_2015.1_eng.pdf;jsessionid=C3EDD415EB9BDC07097AAB2FB707F1B0?sequence=1. Accessed 20 Dec 2019

ZKBS (2018) Synthetic biology. 2nd Interim report of the German Central Committee on Biological Safety, June 2018

4

Synthetic Biology, Biosecurity and Standardisation

Manuel Porcar

The issue of standardisation in Synthetic Biology has important implications at both the technical and governance levels. At the former, standardisation in biology (a still-ongoing process) is expected to exponentially increase the potential of synthetic biology by democratising, easing and expanding our ability to engineer life. Indeed, it has to be stressed that Synthetic Biology is -or at least aims at being- a fully engineering discipline. And engineering, from industrial to electronics, largely relies on standards. A standard is a part, piece, device or procedure with well-established properties, and which can reliably be used in a broad range of industrial applications. Standards are often considered as universal components, in such a way that their constant properties allow a world-wide use. A well-known example of standard parts are nuts and bolts. Indeed, the onset of the industrial revolution was associated with a bloom of different designs of nuts and bolts, with different sizes and thread pitch. It soon became obvious that a standardisation of nuts and bolts was required: standard nuts and bolts were born.

Today, standards are everywhere in our technological civilization. From the screws of our cars to the lids of plastic water bottles, the industrial world we live in would simply be impossible without the use of robust, reliable and standard components. In Synthetic Biology, however, the standardisation process is still in its infancy. Despite the enormous efforts carried out so far to develop and use standard biological parts, plasmids or procedures, the challenge of standardising the biological realm has proved very difficult (Porcar et al. 2015). In this section, we will address the state-of-the-art of standardisation in biology, as well as the implications in biosafety and governance. In order to address this issue, it has first to be stressed than standardisation in the frame of synthetic biology can actually refer to two

M. Porcar (✉)
Institute for Integrative Systems Biology, I2SysBio, University of Valencia, València, Spain
e-mail: manuel.porcar@uv.es

separate and different perspectives, that, for the sake of clarity, we name from now onwards standards *in* biology or standards *for* biology.

4.1 Standards in Biology

In contrast with the standardisation of the norms and regulations of the biosafety or biosecurity issues of synthetic biology (see standards for biology), standards *in* biology are those that are used *within* the discipline. These include, in a matryoshka-like hierarchy of complexity, the following levels amenable to standardisation: parts (i.e. short genetic sequences), devices (simple combinations of the former), genetic circuits, metabolic pathways, engineered cells and engineered cell consortia (or cell tissues). Beyond all those physical components, standards in biology can also refer to the protocols and techniques used in the discipline, as well as the human practices (manual movements, use of biosafety material, etc.) required to carry out any synbio experiment. Figure 4.1, extracted from the BIOROBOOST project, describes this comprehensive and hierarchical structure, which shows the complexity of the endeavor.

This comprehensive description includes all levels in SynBio that are amenable to standardisation. But the question is, do we really have standards at all those levels? The answer is negative, although it is true that major efforts have been made and are still ongoing in order to develop robust standards. In order to get a glimpse of the state-of-the-art in SynBio, two examples of standard parts can be considered: BioBricks™ and pSEVA plasmids.

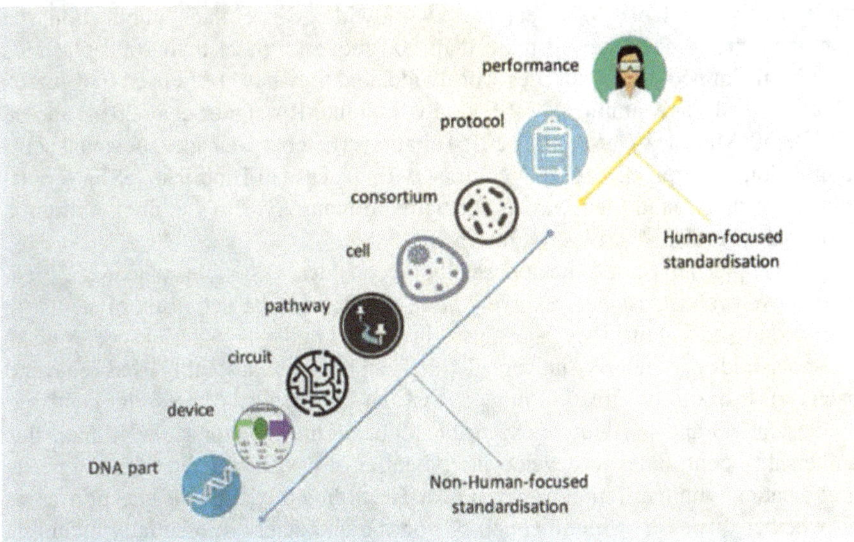

Fig. 4.1 Standards in biology. From DNA part to the protocols and activities of the research staff, all levels are amenable to standardization. (Source: BIOROBOOST project website, http://standardsinsynbio.eu)

BioBricks are DNA sequences which conform to a restriction-enzyme assembly standard, used in SynBio (Smolke 2009). Biobricks were one of the former attempts to standardise engineering in biology, and they are certainly one of the most widely used components, with many scientific references reporting their use, advantages or deficiencies. The main reason behind this wide use is due to the international Genetically Engineered Machine (iGEM) competition (http://www.igem.org), in which hundreds of teams of mainly undergraduate students develop and present ingenuous SynBio projects for which the use of BioBrick parts is imperative. Indeed, iGEM teams not only can use the thousands of BioBricks parts available in the Repository of Biological Parts (http://parts.igem.org/Main_Page) but are requested to provide at least one new part to the registry, in a potent and effective strategy to both broaden the number of available/registered parts and to foster the use of existing Biobricks. Unfortunately, the wide use of Biobricks has not resulted in broad acceptance outside the iGEM world. Moreover, a close analysis of the use of Biobrick-based standard parts by iGEM award-winning teams clearly shows that those teams tend to develop their own standards rather than using those already present in the Registry (Vilanova and Porcar 2014). This lack of trust in the reliability of BioBricks leads to the surprising result that re-inventing the wheel, rather than using standards, is the path to win a standard-centered competition (Fig. 4.2).

Another, very different, example of a widely used standard part in SynBio is that of Standard European Vector Architecture (SEVA) plasmids. SEVA consists of a set of shuttle plasmids developed by the team of Victor de Lorenzo (CNB, CSIC, Madrid, Spain). While most plasmids are only usable in a given bacterial species, shuttle plasmids in particular are functional in a relatively wide range of bacterial species. This means that those plasmids can be used, exchanged and transferred within and among many bacterial species. SEVA plasmids (pSEVA) have been made freely available to researchers worldwide. Since their publication, de Lorenzo's group has received more than 500 requests; more than 2000 plasmids have been shipped to 35 countries worldwide, and they have been used in research which yielded, for example, 277 (SEVA 1.0, Durante-Rodríguez et al. 2014) and 88 (SEVA 2.0, Martínez-García et al. 2015) citations (Esteban Martínez, personal communication). Interestingly, and in contrast with the case of Biobricks, pSEVA plasmids have thus found their way as standards in many SynBio's toolbox without a regulatory (i.e. iGEM's rules) requirement.

pSEVA plasmids could metaphorically be considered "double standards". First, their proven robustness and relatively wide use fit with the definition of standard; and second, the fact that they can be used in a broad range of bacterial species make these plasmids particularly "universal" (they can be interchangeably used in several bacteria). It has to be stressed, though, that the universality of bacterial hosts has both technical and biosafety/biosecurity implications. Regarding the former, this universality contributes to develop the potential of SynBio by facilitating genetic modification within different bacterial taxa. Regarding the latter, the question arises on whether this ease of modification of non-model bacteria (potentially including pathogenic ones) may contribute to an increased concern in terms of the biosecurity and biosafety of those and other technologies being enabled to trespass the "species barrier."

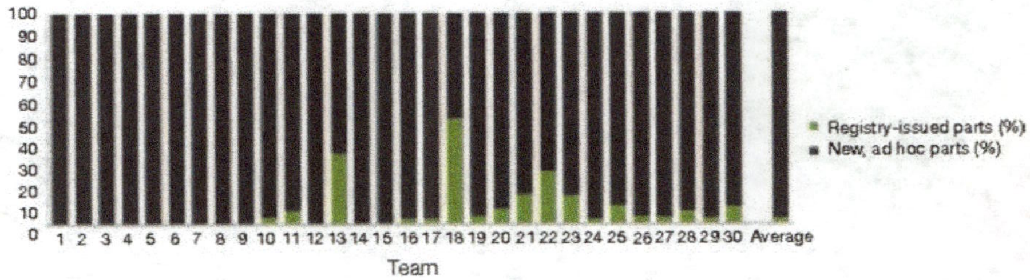

Fig. 4.2 Use of standards in a SynBio standards competition. The figure shows the number of parts from the Registry of Standard Biological parts used in the international Genetically Engineered Machine (iGEM) competition by 30 award-winning teams. An overwhelming number of parts are new ones/unverified and thus not standard. Original figure published in Nature Microbiology and available at https://www.nature.com/articles/nbt.2899

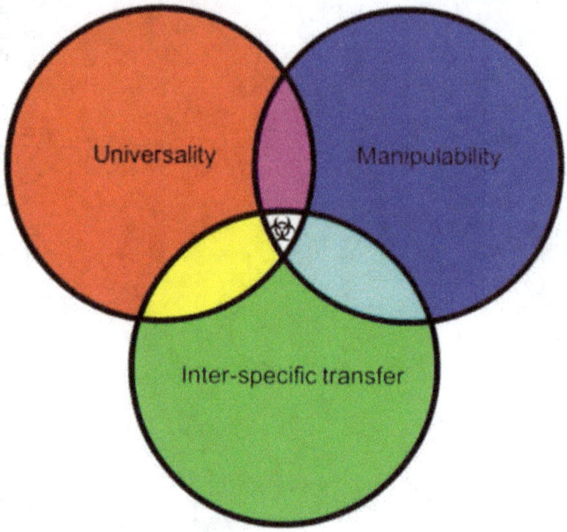

Fig. 4.3 Aspects of biological standardization as they relate to biosecurity

4.2 Implications of Standards in Biosecurity in Terms of Risks

The issue of biological standardization as it relates to biosecurity has not previously been addressed in detail. In the present chapter, we identify a series of aspects linked to standardization and their implications in biosecurity (Fig. 4.3).

4.3 Universality

An example of an almost-universal device is the smartphone. There are millions of them on Earth, and in many countries, most citizens have at least one. Smartphones are standard in the sense that, despite the existence of different models (or strains/species, in biology), they work in an equivalent way. Receiving or sending a Whatsapp message, for example, is largely independent of the smartphone used, because they all work alike with the app. Additionally, an informatic virus, a particular fake new or a geolocation involving smartphones could have an effect on all of them. In other words, the universality of a device is linked to the universality of the risk. Not unlike smartphones, making a standardized platform for SynBio would universalize the risk. If a given plasmid, virus or cellular chassis were made universally available, so would be the risk derived from malicious use.

4.4 Chassis and Trojan Horses

In Homer's Iliad, Greek soldiers entered the city of Troy hidden inside a wooden horse. It must be stressed that the horse was not the weapon, but the vehicle of the actual weapon (the army). Considerable effort was required to set in place the horse as a chassis of the weapon, but once in place, its further use because much easier (although there are no mentions in the Iliad of a further use of the horse). In the example above, smartphones were described as standard devices that may serves as chassis/Trojan horses. Biological chassis, provided that they are robust, easy to maintain and to amenable to modification, could also be considered as biological Trojan horses: inoffensive by themselves, but susceptible of being use as delivers of bioterrorist actions because of their manipulability.

4.5 Breaking Down the Species Barrier

As we have stressed in the previous section, several currently ongoing efforts are successfully allowing microbial transformation by introducing plasmids in a range of different species (see the description of pSEVA above). The obvious implication in terms of biosecurity is that pathogenic DNA fragments could be inserted into harmless bacteria turning them pathogenic or, alternatively, pathogenic bacteria could be turned into more lethal agents by including certain biological circuits from taxonomically distant bacterial species.

4.6 Standards as Social Constructs

As a final remark, we strongly believe that it is important to be aware of a common misconception on the "inner" nature of standards. Robustness, reliability and ease of use are highly relevant features of any standard. That said, though, a standard must be acknowledged within a group of individuals on a basically arbitrary basis (see metric units, flag colours, and any other "conventional" standards). This has implications in terms of biosecurity assessment, since discussions tend to focus on the risk of biological parts *per se*, and not on the risk of standards *because they are standards*.

4.7 Final Remarks and Open Questions

As we have seen in this chapter, standardization in biology is a complex, still in process path that will be central for SynBio to fully develop its potential. Standardisation could finally make SynBio's promise come true and make biology easier to engineer. As we have described above, this fact will ineluctably be linked to an increased risk in the discipline because of the universality of the biological systems (and actors), their amenability as Trojan horses, and the possibility of easily breaking the species barrier. The question arising here is not thus whether advances in standardization will be linked to increased bioterrorism concerns, but to which extent the risk is proportional to the standardization level accomplished. This question does not only affect biosecurity but also biosafety. Consider, for example, the much needed biocontainment of potentially dangerous biological agents: are biocontainment strategies different in a standard-free vs. a standard-complete scenario?

As a general conclusion, the standardization of SynBio is a complex process, mostly still in its infancy. The success of this process will result in immense economic and societal benefits. However, the risks of SynBio in terms of biosecurity are only partially known, and the implications of the possible success of the ongoing standardization process in the biosecurity threads of this emerging discipline deserve further study.

Acknowledgments The author (MP) received funds from the European Union through the BioRoboost project, H2020-NMBP-TR-IND-2018-2020/BIOTEC-01-2018 (CSA), Project ID 210491758.

References

Durante-Rodríguez G, de Lorenzo V, Martínez-García E (2014) The Standard European Vector Architecture (SEVA) plasmid toolkit. Methods Mol Biol 1149:469–478. https://doi.org/10.1007/978-1-4939-0473-0_36.

Martínez-García E, Aparicio T, Goñi-Moreno A, Fraile S, de Lorenzo V (2015) SEVA 2.0: an update of the Standard European Vector Architecture for de-/re-construction of bacterial functionalities. Nucleic Acids Res 43(Database issue):D1183–D1189. https://doi.org/10.1093/nar/gku1114

Porcar M, Danchin A, de Lorenzo V (2015) Confidence, tolerance, and allowance in biological engineering: the nuts and bolts of living things. BioEssays 37(1):95–102. https://doi.org/10.1002/bies.201400091

Smolke CD (2009) Building outside of the box: iGEM and the biobricks foundation. Nat Biotechnol 27(12):1099–1102. https://doi.org/10.1038/nbt1209-1099

Vilanova C, Porcar M (2014) iGEM 2.0 – refoundations for engineering biology. Nat Biotechnol 32(5):420–424. https://doi.org/10.1038/nbt.2899

5

Dual Use Life Sciences and Biosecurity: Responsible Governance

Ineke Malsch and Maria Espona

Abstract This chapter analyses a case study of responsible governance of dual use life sciences and biosecurity in Armenia. It is based on materials presented during a Responsible Research and Innovation Course in Armenia, held on 17–19 May 2019. The course was organised as part of the ISTC Targeted Initiative (TI) on CBRN Export Control and Dual Use in Central Asia. The focus of the case study is on ethical aspects and how collective responsibility for biosecurity can be organised, in order to prevent innovation from undermining international law prohibiting hostile uses of life sciences.

5.1 Introduction

For the purpose of this chapter, dual use technology is defined as civil technology with military or criminal misuse potential. Dual use sciences and technologies are problematic, because banning them is impossible since this would inhibit the development of beneficial peaceful applications such as new medicine or novel food. In addition, the current legal framework is insufficient to govern emerging new technologies. Innovation in dual use technologies calls for additional voluntary self-governance of the scientific and industrial communities concerned.

In general, innovation impacts laws through the following mechanism. Laws governing technologies are formal instruments, adopted by legislative authorities, which regulate the market access and permitted properties and uses of known technological products, processes and systems. Future technological innovation can be addressed to some extent during the formulation of laws, e.g. through "general purpose criteria" in the Chemical and Biological Weapons Conventions prohibiting any kind of chemical, biological or toxin substance for hostile purposes, implicitly covering also substances which did not exist at the time of adopting the convention.

I. Malsch (✉) · M. Espona
Malsch TechnoValuation, Utrecht, The Netherlands
e-mail: malschtechnovaluation@xs4all.nl

However, innovation tends to create uncertain and unforeseen impacts, which often go beyond the imagination of lawmakers (Trump et al. 2020a). This unpredictability of emerging technologies calls for regular updating of existing laws or even adoption of new laws. E.g. the increasing popularity of innovative drones among hobbyists (e.g. airborne cameras) and commercial service providers (e.g. delivery drones) has led to discussion on the adoption of new laws regulating unmanned aerial vehicles (c.f. Finger et al. 2015). Nevertheless, the laws update process is difficult and takes time to catch up with technological innovation, because of that, in many cases, when the new law is out, it is already outdated.

The formulation of legal texts related to science and innovation are based on sound scientific evidence, the collection of which is time consuming, and on consultations with the relevant stakeholders, often with conflicting interests and diverging views. In these circumstances, soft law (e.g. EU or UNESCO Recommendations) or voluntary codes of conduct are often used as complementary instruments to fill the formal legislative deficit.

After presenting the current legal framework governing dual use life sciences and biosecurity, we will discuss some ethical concepts and instruments that could be used to govern this domain. How these ethical instruments can work out in practice, and which stakeholders should be involved is illustrated in the subsequent case study of dual use life sciences in Armenia. The chapter ends with some conclusions and recommendations.

5.2 Capacity Building in Dual Use Export Control in TI Partner Countries

The International Science and Technology Center (ISTC) and the Science and Technology Center in Ukraine (STCU) are intergovernmental organizations established in 1992 and 1993, respectively as unique multilateral mechanisms to prevent proliferation of ex-Soviet scientific know-how related to weapons of mass destruction (WMD). By sponsoring and facilitating scientific cooperation, the two Centers seek to integrate these scientists into the international community and redirect their talents into peaceful, sustainable, civilian work. Between them, the ISTC and STCU Recipient membership include most of the former Soviet Independent States. Both Centers have the legal status of intergovernmental organizations with diplomatic privileges in the member countries where they work.

In the last two decades, the threat that non-state actors may acquire chemical, biological, radioactive or nuclear (CBRN) weapons has become an increasing concern for the international community. Taking into consideration the fast pace of scientific and technological change that is taking place in the ISTC/STCU countries, both Centers have developed new delivery mechanisms for science & technology projects over the past 10 years described as Targeted Initiatives (TI's). One of them is a Targeted Initiative on CBRN Export Control on Dual-Use Materials and Technologies funded by the European Commission.

The partner counties (PC) of the TI are Afghanistan, Armenia, Azerbaijan, Georgia, Kazakhstan, Kyrgyzstan, Moldova, Mongolia, Pakistan, Tajikistan, Turkmenistan, Ukraine, and Uzbekistan.

The Targeted Initiative focuses on mitigating the risks of misuse of the investigations and know-how related to CBRN weapons programs; the raising of awareness among both the academic and industrial community of these risks, as well as facilitating the collaboration between state and non-state actors of Export Control systems in PC's.

The Targeted Initiative includes the following activities:

- Development of a network of scientists and activities related to responsible science and ethics;
- Development of a master course and other customized courses on export control;
- Grants for PhD students from the Partner countries; and
- Outreach to Industry, including handbooks development and commodity identification courses.

Since the start of the activities under the TI, the Centers have carried out several regional seminars to raise awareness among researchers, academics, industry, and government representatives.

In September 2019, in conjunction with Taras Shevchenko National University of Kyiv, Ukraine, a master program on Economic security of Entrepreneurship including modules on export control was launched.

Furthermore, within the TI activities, local experts from Kyrgyzstan, Georgia, Moldova, Ukraine and Azerbaijan developed handbooks on export control. These handbooks will be available free of charge to the public with an intention to facilitate the access to the information and improve the control of dual-use goods.

In Kazakhstan, a group of experts is working on an Internal Compliance Program (ICP) model to be used by industries, which will then be made available to all partner countries.

In August 2019, the Centers held a meeting in Ypres, Belgium, for young scientists, in order to raise awareness of the potential misuse of their research and the importance of ethics in education. The outcome was very impressive with many of the scientists developing projects on ethics and education to be run in their own countries, for young students.

5.3 Legal Framework Governing Dual Use Life Sciences and Biosecurity

There are several legal instruments governing dual use life sciences, each of them with different characteristics and imposing different obligations to signatories and partner countries.

5.3.1 Protocol for the Prohibition of the Use of Asphyxiating, Poisonous or Other Gases, and of Bacteriological Methods of Warfare – Geneva Protocol of 17 June 1925[1]

The Geneva Protocol was drawn up and signed at a conference which was held in Geneva under the auspices of the League of Nations from 4 May to 17 June 1925, and it entered into force on 8 February 1928.

The 1925 Geneva Protocol prohibits the use of chemical and biological weapons in war: "asphyxiating, poisonous, or other gases, and of bacteriological methods of warfare." It recognizes the significance of bringing together controls on chemical and biological weapons. It prohibits the **use** of such weapons. A number of countries submitted reservations when becoming parties to the Geneva Protocol, declaring that they only regarded the non-use obligations as applying to other parties and that these obligations would cease to apply if the prohibited weapons were used against them. The main elements of the protocol are now considered by many to be part of customary international law.

Armenia signed the Protocol on 13 March 2018.

5.3.2 The Convention on the Prohibition of the Development, Production and Stockpiling of Bacteriological (Biological) and Toxin Weapons and on Their Destruction[2]

The Convention on the Prohibition of the Development, Production and Stockpiling of Bacteriological (Biological) and Toxin Weapons and on their Destruction, commonly known as the Biological Weapons Convention (BWC) or Biological and Toxin Weapons Convention (BTWC) is a legally binding multilateral treaty that outlaws biological arms banning an entire category of weapons. The BTWC opened

---

[1] https://unoda-web.s3-accelerate.amazonaws.com/wp-content/uploads/assets/WMD/Bio/pdf/Status_Protocol.pdf

[2] http://disarmament.un.org/treaties/t/bwc/text

for signature in 1972 and entered into force in 1975. It currently has 183 states-parties, including Palestine, and four signatories (Egypt, Haiti, Somalia and Syria). Ten states have neither signed nor ratified the BTWC (Chad, Comoros, Djibouti, Eritrea, Israel, Kiribati, Micronesia, Namibia, South Sudan and Tuvalu).

The BTWC bans the development, production, acquisition, transfer, retention, stockpiling and use of

- Biological agents and toxins "of types and in quantities that have no justification for prophylactic, protective or other peaceful purposes;"
- Weapons, equipment, and delivery vehicles "designed to use such agents or toxins for hostile purposes or in armed conflict."

The convention further requires states-parties to destroy or divert to peaceful purposes the "agents, toxins, weapons, equipment, and means of delivery" described above within nine months of the convention's entry into force. Accordingly, many States Parties have adopted national legislation and regulations to implement the prohibitions of the Convention. The BTWC does not ban the use of biological and toxin weapons but reaffirms the 1925 Geneva Protocol, which prohibits such use. It also does not ban biodefense programs.

In contrast to the Chemical Weapons Convention, the BTWC has no verification mechanism to monitor compliance, and negotiations on the creation of such a mechanism have stalled to date. The treaty regime mandates that states-parties solve compliance concerns consulting each other and also allows states-parties to lodge a complaint with the UN Security Council who can investigate compliance issues; but this power has never been invoked.

At the Sixth Review Conference (2006), the States parties adopted by consensus a detailed plan for promoting universal adherence and decided to update and streamline the procedures for submission and distribution of the Confidence-Building Measures (CBMs). They also adopted a comprehensive intersessional programme spanning from 2007 to 2010 that included activities related to the implementation of the obligations that arise from different articles of the treaty. In a significant development, the Conference agreed to establish an Implementation Support Unit (ISU) to assist States parties in implementing the Convention. The ISU -a 3 persons team-has been established for the BTWC within the Geneva Branch of the United Nations Office for Disarmament Affairs, which is also tasked with outreach activities. But in practice, education, outreach and awareness-raising activities take place in scientific, professional and academic associations, bodies and institutions within States Parties.

Recently, and in particular in the wake of the evolution of more sophisticated/complex terrorism threats and actions, a renewed interest in ensuring greater global participation and implementation of the BTWC has rapidly emerged. So, States Parties agreed to promote the effective implementation of the Convention nationally, including integration into education; outreach; and raising awareness.

In 1994, Armenia joined the Biological and Toxin Weapons Convention (BWC).

5.3.3 United Nation Security Council Resolution No. 1540 (UNSCR 1540)[3]

United Nations Security Council Resolution (UNSCR) 1540 places an international obligation on all Members of the UN to take action against the proliferation of WMD. UNSCR 1540 is a legally binding international instrument with the objective to prevent non-state actors from acquiring and using WMD, which has been widely used as a common legal basis for states to use when drafting and promoting strategic trade control legislation. The resolution also encourages international cooperation to achieve its goals and raise awareness among the national stakeholders to have a sound implementation of the Resolution.

The control lists agreed in the multilateral regimes (Nuclear Suppliers Group, Australia Group, Missile Technology Control Regime and Wassenaar Arrangement) have become an informal part of international law, since they are (vaguely) referred to in UNSCR 1540. In a footnote, UNSCR 1540 defines related materials as 'materials, equipment and technology covered by relevant multilateral treaties and arrangements, or included on national control lists, which could be used for the design, development, production or use of nuclear, chemical and biological weapons and their means of delivery'.

Armenia sent information to the 1540 Committee on the following dates: 9 November 2004, 21 December 2005, 12 December 2007, 4 March 2014 and 22 August 2016.

The country matrix was approved on 23 December 2015. It contains all the relevant provisions related to the implementation of the UNSCR 1540 on the nuclear, chemical and biological fields included in the Constitution, Criminal Code and other legal instruments.

> The 1540 Committee uses the matrices as a reference tool for facilitating technical assistance and to enable the Committee to continue to enhance its dialogue with States on their implementation of Security Council resolution 1540 (2004). The matrices are not a tool for measuring compliance of States in their non-proliferation obligations but for facilitating the implementation of Security Council Resolutions 1540 (2004), 1673 (2006), 1810 (2008), 1977 (2011) and 2325 (2016).[4]

5.3.4 National Laws and Law Enforcement

Internally, each country must fulfil the obligations assumed at the time they become part of the international instruments above-mentioned. According with their legal system, they will issue the laws that are required to comply with the treaties.

---

[3] https://undocs.org/S/RES/1540(2004)

[4] https://www.un.org/en/sc/1540/national-implementation/1540-matrices.shtml

When it comes to the enforcement there are many nuances related to both the extent of the dual use related sector in the country (not only the industry but the universities and academia), the commitment of the state with WMD non-proliferation, the risk perception analysis on the topic and the availability of resources (human and financial) to do the task.

Because of the disparity between countries, many international programs are in place to help them to have a sound implementation of their international obligations.

Our TI on Export Control is linked to the countries' obligations under the UNSC 1540 as well as the BTWC.

In 2010 Armenia adopted a Law on control over the circulation of goods and technologies of dual use which substituted an analogous law adopted in 2003. For the implementation of the law, the Government of Armenia adopted a Decree No 1785 of 15.12.2011, which confirmed the control list of dual-use goods. The list follows the European list of controlled dual-use goods and technologies and it is frequently updated according to the changes of the European lists.

The violations of the obligation of the international treaties regarding biological weapons is included in the Armenian Criminal Code, mainly in articles 284–286 and 386–387.[5]

5.4 Ethics of Dual Use Life Sciences

The legal framework described above constitutes a solid foundation for governing dual use life sciences. As mentioned in the introduction, this legal basis should be complemented by respect for fundamental ethical principles. From an ethical perspective, four concepts are relevant for governance of dual use life sciences: human rights, justice, war and peace, and responsibility.

Human rights are fundamental and inalienable, meaning that the rights of any human being should always be respected, even if they would not be formally written down in a positive legal text. Human rights are universal, meaning that everyone is entitled to the rights. International law includes positive formulations of human rights, distinguishing civil and political liberties, and socio-economic rights.[6] This formalisation is a continuing process leading to the recognition of new rights under the influence of social as well as technological innovations. Some current discussions address the recognition of rights for human embryos, animals and cyborgs (cybernetic organisms, e.g. combining robotic and human body-parts). Several technologies can impact human rights in different ways. For example, innovation in biomedical technologies can affect the right to life, human dignity, informed consent and other health-related rights. Emerging Information and Communication

[5] https://www.unodc.org/res/cld/document/armenia_criminal_code_html/Armenia_Criminal_Code_of_the_Republic_of_Armenia_2009.pdf

[6] C.f. UN Treaty collection: https://treaties.un.org/

Technologies (ICT) influence the way privacy, personal data protection, civil and political rights and the right to decent working conditions can be protected. Novel agricultural or food technologies can affect socio-economic rights of farmers and consumers. Human rights are not absolute: trade-offs between different rights must be found in a continuing balancing act, e.g. between academic freedom for life scientists and biosecurity for society at large.[7]

Justice is a concept which intrinsically calls for balancing the rights of different individuals and social communities. New technology tends to impact this balance in different ways. For example, the right to ownership presupposes a delicate balance between protection of intellectual property for inventors and companies versus the universal right to reap the fruits of progress in science for the common good. As new technologies emerge, this balance may shift. For example, the European Group on Ethics highlighted potential ethical issues related to protection of intellectual property raised by innovation in nanomedicine: "According to the current regulatory system for patenting, some exemptions are allowed with regard to the patentability of therapeutic and surgical procedures. The exemptions in the present patent system are based on a balance of interests whereby diagnosis, therapy and research should be available to patients without patents being a hindrance. This is likely to be blurred because the new nanomaterials may logically fall within more than one category. To protect the ethical position that has led to these exemptions it is important to ensure that patents in these new areas do not alter the current balance. There are risks of overly broad patents being granted that may hinder their therapeutic availability. This is also the case for nanomedicine." (EGE 2007). In addition, fairness may be affected if technology gives rise to a changing gap between haves and have-nots. New surveillance technologies can facilitate shifting the balance between liberty and security towards securitization by "big brothers" (governments monitoring the movements of citizens without a reasonable ground) and "little sisters" (private companies and citizens spying on each other).[8]

Dual use technologies sit squarely at the crossroads between war and peace, which are the focus of several philosophical ethical theories including pacifism[9] and the Just War Theory.[10] The former prohibits the use of violence to solve conflicts between states, while the latter imposes threshold for political decisions to go to war (Jus ad Bellum) and humanitarian rules for conduct of armed forces during a war (Jus in Bello). International Humanitarian Law[11] can be considered a positive formulation of Just War Theory. Some conventions including the Geneva Conventions and arms control treaties including the Biological and Chemical Weapons Conventions extend their scope to military and dual use research and innovation.

---

[7] Read more about the philosophical concept "Human Rights" here: https://plato.stanford.edu/entries/rights-human/

[8] Read more about the philosophical concept "Justice" here: https://plato.stanford.edu/entries/justice/

[9] C.f. https://plato.stanford.edu/entries/pacifism/

[10] C.f. https://plato.stanford.edu/entries/war/

[11] C.f. https://www.icrc.org/en/doc/assets/files/other/what_is_ihl.pdf

Notably, Article 36 of the First Protocol of the Geneva Convention (8 June 1977) reads: "In the study, development, acquisition or adoption of a new weapon, means or method of warfare, a High Contracting Party is under an obligation to determine whether its employment would, in some or all circumstances, be prohibited by this Protocol or by any other rule of international law applicable to the High Contracting Party."

Dual use life sciences are simultaneously subject to human rights including the right to academic freedom and ownership of intellectual property, and to international humanitarian rights, including the ban on use of life science knowledge and technologies for hostile purposes. Malsch (2013) includes a suggestion for balancing these rights, by extending the scope of some Just War principles to military and civil security research and balancing these with academic ethical principles governing peaceful uses. In the early stages of dual use research, from basic research (Technology Readiness Level, TRL 1)[12] until TRL 5 (technology validated in relevant environment) as well as commercial dual use research from TRL 6 (technology demonstrated in relevant environment) until TRL9 (actual system proven in operational environment), care should be taken to balance freedom and security, and dual use risk and technology assessment should be performed in parallel to the R&D. Advance civil security research (TRL 6–9), such as the development of sensor technologies for controlling access to life science laboratories, should in addition address the Just War principles of Just Intent and Proportionality. Advanced legitimate military research (TRL 6–9) such as biodefence is subject only to Just War principles Just Intent, Legitimate Authority and Proportionality.

The principle of collective responsibility for progress in science and technology was formulated by the philosopher Hans Jonas (1979). In short, he asserted that science and technology can potentially contribute to catastrophic risks, e.g. by large scale use of biological, chemical, or nuclear weapons. To prevent the extinction of humankind he foresaw, all people must contribute to a collective responsibility. In his view, worst case scenarios must be developed as the basis for strict regulation of science and technology. Since then, the interdisciplinary field of technology assessment has emerged (c.f. Banta 2009), and several methodologies have been developed for risk management and risk governance. Currently, most experts seek to strike a balance between potential benefits and risks. A related trend is the shift from government to governance. Government implies that national authorities in a State have the sole responsibility to protect their citizens, environment, and national interests. Governance implies that governments share a collective responsibility with non-state actors, including companies, the research community, non-governmental organisations (NGOs) and citizens. However, responsible governance of innovation is not straightforward, because of "many hands" (c.f. Poel et al. 2015) or "wicked problems"[13]: in the current system with a globalised economy, national legal

[12] Read more on Technology Readiness Levels: https://en.wikipedia.org/wiki/Technology_readiness_level

[13] C.f.: https://en.wikipedia.org/wiki/Wicked_problem

sovereignty and international governmental organisations, too many parties take responsibility for their own part of the problem, nobody has the capacity to take responsible for the common good, and unforeseen negative consequences are not addressed (Trump et al. 2020a, b). The need for institutional reform is often stressed by UN institutions, Multinational companies (World Economic Forum), International NGOs and the Scientific community (c.f. Malsch 2018).

5.5 Case Study Life Sciences and Biosecurity in Armenia

On 17–19 May 2019, we organised a Responsible Research and Innovation Course for young researchers in Armenia, together with Kai Ilchmann and Hrayr Azizbekyan in the framework of the ISTC TI on export control. An interdisciplinary group of 15 young researchers participated. One of the key case studies demonstrated how innovation can undermine law focused on biosecurity issues. The present chapter discusses the responsible governance of dual use life sciences and biosecurity in Armenia, based on the course materials.

Biology, biomedical sciences, agriculture and other life sciences are double edged swords. The same knowledge, materials and technologies are needed to cure or feed people but can also be misused to kill people or to destroy livestock or food crops (biological weapons). Bio risk management is needed, consisting of two distinct approached. Biosafety aims to "protect people from bad bugs" (natural or accidental infectious disease outbreaks), while biosecurity aims to "protect bugs from bad people" (human-made epidemics). Biosecurity can be protected through legal, social, and technological means. In the legal dimension, biological weapons are prohibited through the abovementioned international agreements and national laws, but also through professional codes of conduct for researchers and companies, including the Statement on Biosecurity published by the Inter-Academies Partnership (IAP) in 2005.[14] The code includes these principles: Awareness, Safety and Security, Education and Information, Accountability and Oversight. The National Academy of Sciences of Armenia is among the 70 national academies which endorsed this code.

Technological biosecurity measures include infrastructure and equipment in laboratories at increasingly stringent biosafety levels (1–4).[15] In addition to technological means for protecting the Occupational Health and Safety of laboratory staff, safer-by-design methods are under development, which can limit the feasibility of misusing life sciences for hostile purposes. Social biosecurity measures include peer review of laboratory risk management including site visits. Researcher training, security screening of employees and dialogue with stakeholders and citizens

---

[14] https://www.interacademies.org/17806/Biosecurity and http://www.interacademies.org/10878/13912.aspx

[15] https://en.wikipedia.org/wiki/Biosafety_level

about biosecurity measures are also useful social means to raise awareness of biosecurity issues.

5.6 Dual Use Life Sciences in Armenia

Life science research is performed in Yerevan State University, Armenian State Agrarian University, and several institutes of the National Academy of Sciences: the institutes of Zoology, Hydroponics, Botany, Molecular Biology, Organic Chemistry and the centre of microbiology and deposition of microorganisms. The Scientific Centres of Agrobiotechnology and of Vegetable-Melons and Industrial Crops of the ministry of agriculture and the Scientific Research Institute of Biotechnology of the Ministry of Economy are also involved (Hovhannisyan and Yesayan 2010). In addition to the biological research in universities, nature is also a possible source of biosafety and biosecurity risks. Some infectious diseases are endemic in the territory of Armenia, including plague (in 80% of the country), tularaemia, anthrax, yersiniosis, leptospirosis, and erysipeloid (Hovhannisyan and Yesayan 2010).

As mentioned in the section on the regulatory framework, Armenia has adopted the BWC already in 1994. The capacity for bio risk management has been built up in international projects since then. Recent international cooperation projects include the following. In 2013–15, UNICRI[16] funded several biosecurity projects in Armenia, focusing on food safety risk assessment, and the work of ministries for emergencies, agriculture, public health, and livestock disease surveillance. The USA and EU have also invested in infrastructural laboratory safety and building human capacity to comply with biosafety protocols including the WHO biosafety manual. Armenian researchers including Avetisyan et al. (2017) surveyed the status quo in medical laboratories and recommended introducing biosafety manuals and training for the personnel. Likewise, Danielyan and Mnatsakanyan (2017) recommended setting up training, working groups and investment in biosafety and biosecurity measures. Indeed, such training and capacity building has continued since then. For example, in 2018, the International Science and Technology Centre (ISTC) organised seven workshops in Armenia to strengthen laboratory biosafety, training 75 participants from 38 laboratories, with EU funding. In addition, ISTC organised a seminar on dual use export control in Yerevan.[17]

[16] http://www.unicri.it/index.php

[17] C.f. www.istc.int

5.7 Discussing Ethical Dilemmas

While most of the capacity building on dual use export control focused on transferring knowledge on national and international legal requirements and on technological and infrastructural safeguards for implementing biosecurity and biosafety, in the Responsible Research and Innovation Course, we raised some ethical dilemmas connected to dual use life sciences and biosecurity issues. The first addressed what would be an appropriate balance between openness and confidentiality in academic dual use research. The second addressed balancing public and private responsibilities in governing biosecurity. The third addressed the acceptability of do-it-yourself biology with dual use potential. All three cases were based on ethical issues which had been discussed in other countries, where private biotechnology companies and do-it-yourself biology laboratories were more common than in Armenia. For example, some participating young researchers had not heard about do-it-yourself biology and were concerned about the implications of allowing this outside of the institutional oversight in universities.

As most of the dual use life sciences research in Armenia is academic at Technology Readiness Levels 1–5, Malsch (2013) would recommend dialogue to reach agreement on the appropriate balance between fostering academic freedom and protection biosafety and biosecurity. The capacity building activities including training staff in biosafety and biosecurity measures are suitable ways to foster such dialogue. In addition, the research performed in the facilities should be accompanied by dual use risk and technology assessment, to monitor emerging biosafety and biosecurity risks.

During the RRI Course, we pointed out the limitations of legal instruments for governing dual use life sciences and called for collective responsibility for dual use research in a "web of prevention" perspective (Dando 2000; Rappert and McLeish 2007). In this case, regulation is not enough, because lifesaving and economically useful research is inherently dual use. In addition, rapid progress in science and technology undermines existing legislation. Collective responsibility for frontier research implies distinct role responsibilities for the involved actors. Governments should perform oversight, improve regulation, and organise public and stakeholder dialogue. Scientists should contribute their expertise in science for policy, train students, develop and abide by codes of conduct, conduct confidence building peer review, participate in academic and public dialogue, and apply their knowledge in safer-by-design research. Companies handling dual use biological materials including pharmaceutical and food producing industry should take their corporate social responsibility, develop and abide by biosafety and biosecurity codes of conduct, inform policy makers about innovations which could influence the current legal framework governing biosecurity, host site visits as part of confidence building measures, and participate in public and stakeholder dialogue. NGOs should contribute to raising public awareness of biosecurity issues, campaign for governmental oversight of dual use life sciences activities, and participate in public dialogue. Citizens should also participate in public dialogue.

5.8 Conclusions and Recommendations

Besides knowledge transfer, the course programme featured lively discussions about the role of scientists in the contemporary Armenian society. Several participants were keen on developing skills in science communication. Inspired by existing codes of conduct including the IAP Statement on Biosecurity, the participants expressed the intention to develop an "Apaga Declaration" including common ethical principles governing their research. The development of such a new contextualised declaration is more important than simply endorsing an existing code because the underlying rationale is to raise awareness about ethical issues and to stimulate scientists to reflect on the decisions they take in their daily practice. In addition to investing in infrastructural and institutional conditions ensuring biosafety and biosecurity, it is important to continuously train young generations of scientists in laboratory safety practices. In addition, these young generations will need to acquire skills in two-way science communication, to be able to participate in public debate around responsible science and ethics topics. Civil Society Organisations and communities are becoming more critical of the role of science in society, and it is important that scientists are not only persuasive in transmitting their enthusiasm for science, but also aware of genuine concerns about potential risks and ethical issues related to their research.

Acknowledgement We gratefully acknowledge that the RRIC course was organised with support of the ISTC. The content and format of this chapter are the sole responsibility of the authors.

References

Avetisyan A (2017) Evaluating laboratory biosafety in #1 branch laboratory of NCDCP in Yerevan, Armenia. Poster presentation at 20th Annual Conference of the European Biosafety Association, 25–28 April 2017, Madrid, Spain

Banta D (2009) What is technology assessment? Int J Technol Assess Health Care 25(Suppl 1):7–9. https://doi.org/10.1017/S0266462309090333. https://www.ncbi.nlm.nih.gov/pubmed/19519979

Biological and Toxin Weapons Convention (1975). www.unog.ch/bwc

Dando MR (2000) The new biological weapons: threat, proliferation and control. Lynne Riener Publishers, Boulder

Danielyan A, Mnatsakanya D (2017) Biosafety and Biosecurity Assessment in Reference Laboratory Centre of NCDC Armenia. Poster Presentation at EBSA 2017 proceedings: https://ebsaweb.eu/

EGE (2007) Opinion on the ethical aspects of nanomedicine. opinion no. 21. European Commission, Brussels. https://ec.europa.eu/info/research-and-innovation/strategy/support-policy-making/scientific-support-eu-policies/ege_en

Finger M, Bert N, Kupfer D (eds) (2015) Regulating drones – creating European Regulation that is smart and proportionate. https://cadmus.eui.eu/bitstream/handle/1814/38843/ETR_Observer_2015_03.pdf

Geneva Convention Article 36 – First Protocol (8 June 1977)

Geneva Protocol for the Prohibition of the Use in War of Asphyxiating, Poisonous or Other Gases, and of Bacteriological Methods of Warfare (1925)

Hovhannisyan N, Yesayan A (2010) Becurity and biosafety in developing countries. An example of Republic of Armenia. Presentation at Annual International Symposium - Biosecurity and Biosafety: Future Trends and Solutions. Milan October 13–15, 2010

IAP Biosecurity Working Group since 2003. http://www.interacademies.org/BiotechnologyandBiosecurity.aspx

Jonas H (1979) Das Prinzip Verantwortung. Versuch einer Ethik für die technologische Zivilisation. The principle of responsibility, Frankfürt/M

Malsch I (2013) The just war theory and the ethical governance of research. Sci Eng Ethics 19(2)

Malsch I (2018) Future technologies we want. Wolf Publishers

NTI, compliance with WMD conventions – Armenia. https://www.nti.org/learn/countries/armenia/

Rappert B, McLeish C (eds) (2007, 2012) A web of prevention. Biological weapons, life sciences, and the governance of research. Routledge

Security Council resolution 1540 (2004). https://www.un.org/disarmament/wmd/sc1540/

Trump BD, Cummings CL, Kuzma J, Linkov I (2020a) Synthetic biology 2020: Frontiers in risk analysis and governance. Springer

Trump BD, Galaitsi SE, Appleton E, Bleijs DA, Florin MV, Gollihar JD et al (2020b) Building biosecurity for synthetic biology. Mol Syst Biol 16(7):e9723

UNICRI, CBRN projects Armenia. http://www.unicri.it/topics/cbrn/biosecurity/se_europe_and_caucasus/

van de Poel I, Royakkers L, Zwart SD (2015) Moral responsibility and the problem of many hands. Routledge. 226pp, ISBN:9781138838550

6

Bottom-Up Grassroots Governance and Biosecurity Threats

Tatyana Novossiolova, Todd Kuiken, Jared DeCoste, Luc Henry, Ineke Malsch, Myriam Merad, Carl Newman, Wallace Patterson and Alexandra Waskow

6.1 Introduction

In the run-up to the Seventh Review Conference of the Biological and Toxin Weapons Convention (BTWC) – the principal international agreement that outlaws the development, production, stockpiling, acquisition, and retention of biological and toxin weapons – the US National Research Council published a report which

T. Novossiolova (✉)
Law Program, Center for the Study of Democracy, Sofia, Bulgaria
e-mail: tatyana.novossiolova@csd.bg

T. Kuiken
North Carolina State University, Raleigh, NC, USA

J. DeCoste
US Army Chemical Capabilities Development Command Chemical Biological Center (CCDC CBC), Aberdeen Proving Ground, MD, USA

L. Henry
Ecole polytechnique fédérale de Lausanne (EPFL), Lausanne, Switzerland

I. Malsch
Malsch TechnoValuation, Utrecht, The Netherlands

M. Merad
CNRS (French National Centre for Scientific Research), Paris, France

C. Newman
Biological Threat Reduction Program, Defense Threat Reduction Agency (DTRA), Fort Belvoir, VA, USA

W. Patterson
Air Force Research Laboratory (AFRL), Dayton, OH, USA

A. Waskow
Ecole polytechnique fédérale de Lausanne (EPFL), Lausanne, Switzerland

highlighted three trends in science and technology that affect the scope and operation of the BTWC:

- The rapid pace of change in the life sciences and related fields;
- The increasing diffusion of life sciences research capacity and its applications, both geographically and outside traditional research environments; and
- The extent to which scientific and technical (engineering) disciplines beyond biology are increasingly involved in life sciences research and innovation.[1]

The advancement of synthetic biology over the past two decades epitomises these three trends and underscores the need to develop and implement effective mechanisms for safeguarding all life sciences activities against accidental or deliberate misuse.

Synthetic biology is a fast-growing interdisciplinary field that combines the principles of engineering with the knowledge in biology to generate technologies and products with applications in agriculture, healthcare, foods, materials science, and more by enabling the design, redesign, manufacture and/or modification of genetic material, living organisms, and biological (eco)systems.[2] Such enabling capabilities are inherently dual-use: on the one hand, they can benefit the advancement of life science R&D, but on the other, they might also be misused to cause harm to humans, animals, or the environment (e.g. through the development of biological weapons). The biosafety and biosecurity implications of synthetic biology are being considered within the framework of different international agreements, such as the Convention on Biological Diversity (CBD), Chemical Weapons Convention (CWC), the International Health Regulations (IHRs), and the BTWC. In 2014, the Conference of the Parties to the Convention on Biological Diversity, recognising the potential

---

[1] National Research Council, *Life Sciences and Related Fields: Trends Relevant to the Biological Weapons Convention.* Washington, DC: The National Academies Press, 2011, available at https://doi.org/10.17226/13130. See also J.A. Carrera, A.J. Castiglioni, P.M. Heine, 'Chemical and Biological Contract Manufacturing Services: Potential Proliferation Concerns and Impacts on Strategic Trade Controls', *Strategic Trade Review*, Vol. 3:4, Spring 2017. pp 25–46, available at http://www.str.ulg.ac.be/wp-content/uploads/2017/04/Chemical-and-Biological-Contract-Manufacturing-Services-Potential-Proliferation-Concerns-and-Impacts-on-Strategic-Trade-Controls.pdf.

[2] A. Nouri and S. Seyedin-Noor, 'Synthetic Biology: A Call for a New Culture of Responsibility', *Bulletin of Atomic Scientists,* 7 December 2018, available at https://thebulletin.org/2018/12/synthetic-biology-a-call-for-a-new-culture-of-responsibility/; CBD Ad Hoc Technical Expert Group on Synthetic Biology, *Report of the Ad Hoc Technical Expert Group on Synthetic Biology,* UNEP/CBD/SYNBIO/AHTEG/2015/1/3, 7 October 2015, Montreal, Canada, available at https://www.cbd.int/doc/meetings/synbio/synbioahteg-2015-01/official/synbioahteg-2015-01-03-en.pdf; NASEM, *Biodefense in the Age of Synthetic Biology*, National Academies Press, Washington DC, 2018; J. Zhang et al., *The Transnational Governance of Synthetic Biology: Scientific Uncertainty, Cross-Borderness, and the 'Art' of Governance,* BIOS Working Paper No. 4, 2011, BIOS, London School of Economics and Political Science, London. Available at https://royalsociety.org/~/media/Royal_Society_Content/policy/publications/2011/4294977685.pdf; R. Carlson, *Biology is Technology: The Promise, Peril, and New Business of Engineering Life,* Harvard University Press, Cambridge MA, 2010.

impact that technologies with synthetic life, cells or genomes can have on the conservation and sustainable use of biological diversity, decided to establish an Ad Hoc Technical Expert Group on Synthetic Biology.[3] The mandate of the AHTEG includes, *inter alia,* the identification of the potential benefits and risks of organisms, components, and products arising from synthetic biology techniques to the conservation and sustainable use of biodiversity, as well as any related human health and socioeconomic impacts relevant to the Convention and its Protocols.[4] The work of the AHTEG is directly pertinent to the functioning of the Cartagena Protocol on Biosafety to the CBD. The Cartagena Protocol aims to ensure the safe handling, transport, and use of living modified organisms (LMOs) resulting from modern biotechnology that may have adverse effects on biological diversity, taking also into account risks to human health.[5] The convergence of chemistry and biology and its implications for the Chemical Weapons Convention – the principal international agreement that outlaws the development, production, and use of chemical weapons – are being reviewed by the Scientific Advisory Board (SAB) of the Organisation for the Prohibition of Chemical Weapons (OPCW).[6] The benefits and risks of recent life science advances such as genome editing have been considered during the BTWC proceedings, in order to ensure that related knowledge, materials, and techniques are utilised only for peaceful, protective, and prophylactic purposes.[7] Synthetic biology has also been addressed in the context of global health security, particularly with regard to the need for strengthening laboratory biosafety and biosecurity norms and fostering a culture of responsibility in the life sciences.[8]

Despite the growing recognition of the dual-use potential of synthetic biology, developing viable mechanisms for mitigating biosafety and biosecurity concerns

---

[3] Conference of the Parties to the Convention on Biological Diversity, *Decision XII/24. New and emerging issues: synthetic biology,* UNEP/CBD/COP/DEC/XII/24, 17 October 2014, Pyeongchang, Republic of Korea, available at https://www.cbd.int/doc/decisions/cop-12/cop-12-dec-24-en.pdf.

[4] Conference of the Parties to the Convention on Biological Diversity, *Decision XII/24. New and emerging issues: synthetic biology,* UNEP/CBD/COP/DEC/XII/24, 17 October 2014, Pyeongchang, Republic of Korea, available at https://www.cbd.int/doc/decisions/cop-12/cop-12-dec-24-en.pdf.

[5] Cartagena Protocol on Biosafety to the Convention on Biological Diversity, 11 September 2003, available at http://bch.cbd.int/protocol.

[6] OPCW Scientific Advisory Board, *Report of the Scientific Advisory Board on Developments in Science and Technology for the Fourth Special Session of the Conference of the States Parties to Review the Operation of the Chemical Weapons Convention,* RC-4/DG.1, 30 April 2018, The Hague, The Netherlands, available at https://www.opcw.org/sites/default/files/documents/CSP/RC-4/en/rc4dg01_e_.pdf.

[7] See, for example, BTWC Meeting of Experts on Review of developments in the field of science and technology related to the Convention, Report of the 2018 Meeting of Experts on review of developments in the field of science and technology related to the Convention, BWC/MSP/2018/MX.2/3, 12 November 2018, Geneva, Switzerland, available at https://undocs.org/BWC/MSP/2018/MX.2/3.

[8] World Health Organisation, *Responsible Life Sciences Research for Global Health Security: A Guidance Document,* 2010, Geneva, Switzerland, available at https://www.who.int/csr/resources/publications/HSE_GAR_BDP_2010_2/en/.

without significantly impeding research and innovation remains a challenge. This is largely due to the fact that under international law dual-use research *per se* is not illicit, as long as it meets the general purpose criterion enshrined in Article I of the BTWC:

> "Each State Party to this Convention undertakes never in any circumstances to develop, produce, stockpile or otherwise acquire or retain:
> (1) Microbial or other biological agents, or toxins whatever their origin or method of production, of types and quantities that have no justification for prophylactic, protective, or other peaceful purposes. Biological and Toxin Weapons Convention, https://www.un.org/disarmament/biological-weapons/."

The general purpose criterion is intended as a comprehensive prohibition of biological and toxin weapons. However, as the negotiations on the development of an international verification protocol to the BTWC have demonstrated, devising a technical system for compliance with the provisions of the Convention has significant limitations, which, unless fully addressed run the risk of undermining the effectiveness of the biological prohibition regime and compromising its integrity. Upholding the general purpose criterion thus requires an integrated set of policies, initiatives, and measures that are flexible and accommodating of the interests and goals of the different stakeholders. In other words, it requires the in-depth implementation of the existing international biosafety and biosecurity regulations, in order to promote, foster, and sustain a strong and viable culture of biosafety, biosecurity, and responsible conduct of science. Such a culture manifests itself in shared beliefs, attitudes, and patterns of behaviour of individuals and organisations that can support, complement or enhance operating procedures, rules, and practices, as well as professional standards and ethics designed to prevent the unintentional (accidental) or intentional release of biological agents and toxins.[9] A robust safety and security culture is an integral element of high reliability organisations and an essential prerequisite for mitigating the risk of 'normal accidents' associated with advanced technology.[10] The Eighth Review Conference of the BTWC in 2016 acknowledged the essential contribution that the life science community can make to promoting and sustaining such a culture. When considering the national implementation of the Convention, the Conference agreed on the value of measures to:

---

[9] International Working Group, *A Guide to Training and Information Resources on the Culture of Biosafety, Biosecurity, and Responsible Conduct in the Life Sciences,* 2019, available at https://absa.org/wp-content/uploads/2019/04/CULTURE_TRAINING_CATALOGUE.pdf.

[10] K. Roberts, 'New Challenges in Organisational Research: High Reliability Organisations', *Organizations and Environment,* vol. 3:2 (1989), pp. 111–125; G. Rochlin, 'Reliable Organisations: Present Research and Future Directions', *Journal of Contingencies and Crisis Management,* vol. 4:2 (1996), pp. 55–59; Ch. Perrow, *Normal Accidents: Living with High-Risk Technologies* (New Jersey: Princeton University Press, 1999); N. Goodman, *Shifting the Blame: Literature, Law, and the Theory of Accidents in Nineteenth-Century America,* (Princeton, NJ: Princeton University Press, 1998); Tim Trevan 'Biological Research: Rethink Biosafety', *Nature,* 11 November 2015.

(a) implement voluntary management standards on biosafety and biosecurity;
(b) encourage the consideration of development of appropriate arrangements to promote awareness among relevant professionals in the private and public sectors and throughout relevant scientific and administrative activities;
(c) promote amongst those working in the biological sciences awareness of the obligations of States Parties under the Convention, as well as relevant national legislation and guidelines;
(d) promote the development of training and education programmes for those granted access to biological agents and toxins relevant to the Convention and for those with the knowledge or capacity to modify such agents and toxins;
(e) encourage the promotion of a culture of responsibility amongst relevant national professionals and the voluntary development, adoption and promulgation of codes of conduct.[11]

The purpose of this chapter is to examine the role of the synthetic biology community in strengthening biosafety and biosecurity and safeguarding synthetic biology against accidental and deliberate misuse. The chapter argues that biosafety and biosecurity education, awareness-raising, and outreach are essential for fostering effective bottom-up (self-governance) approaches for biosafety and biosecurity risk management. Section 6.2 provides an overview of the structure of the synthetic biology community underscoring its complexity in terms of (1) professional interdisciplinarity, (2) diversity of stakeholders, and (3) dynamic landscape with professional and non-professional actors moving from one context to another over time. Section 6.3 then examines the prevalent perceptions and framing of biosafety and biosecurity risks within the synthetic biology community, in order to identify options for enhancing stakeholder engagement and leveraging the diversity of expertise within the synthetic biology community for promoting responsible research and innovation practices (Sect. 6.4). The conclusion (Sect. 6.5) outlines a summary of the key findings in this chapter.

6.2 Structure of the Synthetic Biology Community

Synthetic biology has developed as a result of the convergence of knowledge, techniques, and tools of different scientific disciplines such as systems biology, genetic engineering, mechanical and electrical engineering, information technology, physics, chemistry, nanotechnologies, and computer modelling.[12] The Lego analogy is commonly used to illustrate the potential of synthetic biology: just as Lego bricks of different colour, shape, and size can be combined together to build new

[11] Eighth Review Conference of the States Parties of the Convention on the Prohibition of the Development, Production and Stockpiling of Bacteriological (Biological) and Toxin Weapons and on Their Destruction, *Final Document,* BWC/CONF.VIII/4, 11 January 2017, Geneva, Switzerland, available at https://undocs.org/BWC/CONF.VIII/4.

[12] United Nations Interregional Crime and Justice Research Institute (UNICRI), *Security Implications of Synthetic Biology and Nanobiotechnology: A Risk and Response Assessment of Advances in Biotechnology* (Turin: UNICRI, 2012), http://www.unicri.it/special_topics/nanobiotechnology/security_report/.

structures, so can genes and proteins be used as building blocks to create new kinds of cells and new biological functions for cells.[13] In some respects, the evolution of synthetic biology follows the consolidation of molecular biology throughout the 1930s, which facilitated the characterisation of the structure of the DNA molecule and culminated in the emergence of gene splicing experiments in the early 1970s. As noted in the National Research Council 2009 report, 'A New Biology for the 21st Century':

> Biology is at a point of inflection. Years of research have generated detailed information about the components of the complex systems that characterize life – genes, cells, organisms, ecosystems – and this knowledge has begun to fuse into greater understanding of how all those components work together as systems. Powerful tools are allowing biologists to probe complex systems in ever-greater detail, from molecular events in individual cells to global biogeochemical cycles. Integration within biology and increasingly fruitful collaboration with physical, earth, and computational scientists, mathematicians and engineers are making it possible to predict and control the activities of biological systems in ever greater detail. [14]

Within this context, synthetic biology could be considered a game-changing technology, rather than just a novel scientific discipline.[15] For one thing, it allows an unprecedented access to cutting-edge tools, techniques, and methods for manipulating biological and biochemical systems to an increasingly diverse range of practitioners outside traditional life science domains. This is manifested in the horizontal interdisciplinary diversity across the synthetic biology community which includes engineers, computer and materials scientists, and chemists. Synthetic biology has further attracted the interest of designers and artists, as well as given rise to a fast-growing global 'do-it-yourself' (DIY) movement of 'amateur/garage' biologists, some having little or no formal science education or research credentials.[16]

The synthetic biology community is heterogeneous in terms of stakeholders, too. As far as academic research and teaching are concerned, relevant courses are embedded within the formal curricula of universities at undergraduate and postgraduate level around the world. There are specialised synthetic biology academic research centres, institutes, and ad-hoc societies for advancing innovation and scholarship. A case in point is the BioBricks Foundation, a not-for-profit organisation set up in 2006, in order to promote the use of standardized biological parts that

[13] J. Collins, 'Synthetic Biology: Bits and Pieces Come to Life', *Nature,* vol. 483 (2012), pp. S8–S10.

[14] National Research Council, *A New Biology for the 21st Century* (Washington DC: National Academies Press, 2009), available at https://www.nap.edu/catalog/12764/a-new-biology-for-the-21st-century.

[15] R. Carlson, Biology is Technology: The Promise, Peril, and New Business of Engineering Life, Harvard University Press, Cambridge MA, 2010.

[16] R. Sleator, 'Synthetic Biology: From Mainstream to Counterculture', *Archives in Microbiology,* vol. 198 (2016), pp. 711–713, https://link.springer.com/article/10.1007/s00203-016-1257-x.

are safe, ethical, cost effective and openly accessible.[17] National science academies, individually and collectively carry out periodic monitoring of the state of science in the field, organise meetings and events, and conduct high-level assessments of the social, economic, environmental, or security impact of novel scientific and technological advances.[18]

Synthetic biology finds a wide-ranging application in the biotechnology industry, particularly in the field of drug development, plant breeding, food production, and as an alternative to petrochemical manufacturing. The commercial sector further includes gene synthesis companies – firms that sell synthetic DNA – as well as start-up companies, social entrepreneurs, and bio-incubators – organisations and spaces that help projects and startups develop into mature and sustainable businesses.[19]

Professional associations within industry and academia play an important role in developing standardised approaches and practices, promoting competence and excellence, and recognising and rewarding positive behaviour. They can also act as interlocutors during policy- and decision-making processes.

R&D in the area of synthetic biology benefits from public and private funding from a variety of sources, including government agencies, private foundations and charities, venture philanthropies, and investors. Government agencies are further involved in the administration and regulation of science and research activities.

Science publishers and mass media, including social networks constitute another critical stakeholder, not least because of their role in shaping public opinion and their responsibility to ensure rigorous and ethical reporting and dissemination of information.

The availability of kits, affordable equipment, and commercial services has facilitated the emergence of community-style laboratories effectively turning the practice of biology into a leisure activity open to individuals from all walks of life.[20] Through their activities, non-traditional actors interested in the life sciences, such as Do-It-Yourself (DIY) biologists, designers, and artists seek to promote a better understanding of biotechnology and ultimately uncover new creative ways of resolving societal challenges.

---

[17] BioBricks Foundation, 2020, available at https://web.archive.org/web/20151113084040/http://biobricks.org/about-foundation/.

[18] See, for example, European Academies Science Advisory Council, *Synthetic Biology: An Introduction,* (Brussels: EASAC, 2011), available at https://easac.eu/publications/details/synthetic-biology-an-introduction/; Organisation for Economic Cooperation and Development and the Royal Society, *Symposium on Opportunities and Challenges in the Emerging Field of Synthetic Biology,* (OECD, Royal Society, 2010), available at https://sites.nationalacademies.org/pga/stl/PGA_050738.

[19] For more information about bio-incubators, see https://sphere.diybio.org/.

[20] National Research Council, *Life Sciences and Related Fields: Trends Relevant to the Biological Weapons Convention*, (Washington, DC: The National Academies Press, 2011), available at https://doi.org/10.17226/13130; L. Scheifele and T. Burkett, 'The First Three Years of a Community Lab: Lessons Learned and Ways Forward', *Journal of Microbiology and Biology Education,* vol. 17:1 (2016), pp. 81–85.

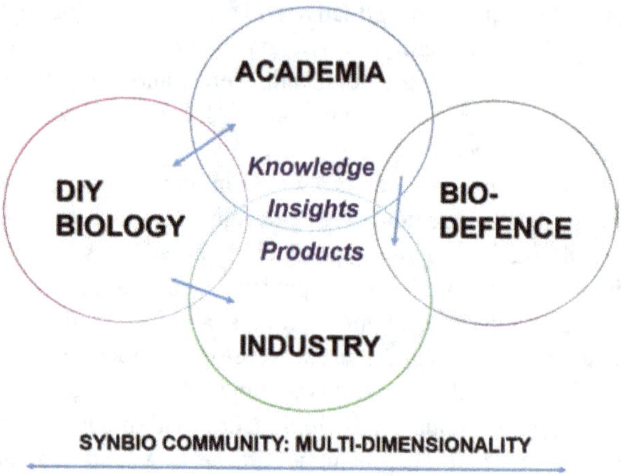

Fig. 6.1 Structure of the synthetic biology professional community (Source: Authors)

The general public is the primary beneficiary and end-user of the materials and products generated as a result of the progress of synthetic biology. Ensuring R&D integrity, safety, security, and reliability are essential requirements for maintaining public trust in science.

The field of synthetic biology is very dynamic, allowing practitioners to frequently change jobs and professional settings. It is possible, for instance, to move from academic research to industry and vice versa; to take up biology as a hobby with the prospect of becoming a social entrepreneur; or to start one's own business during or after formal schooling. This means that professionals often get exposed to different professional cultures, which in turn, enables them to develop a range of transferrable skills and increase their capacity for professional adaptation (Fig. 6.1).

6.3 Perceptions of Risks Within the Synthetic Biology Community

Risks associated with synthetic biology generally fall into two overarching categories: biosafety risks that result from accidents or negligent behaviour; and biosecurity risks that result from the deliberate misuse of knowledge, information, or materials. The term 'biosafety' is defined differently by stakeholders. For the purposes of the present chapter, two definitions of 'biosafety' are considered, namely the definition of the World Health Organisation and the definition accepted under the Convention on Biological Diversity. The World Health Organisation defines biosafety as the set of 'containment principles, technologies, and practices that are implemented to prevent the unintentional exposure to biological agents or their inadvertent release'.[21] Within the context of the CBD, 'biosafety' is understood as the 'safe handling, transport and use of living modified organisms (LMOs) resulting from modern biotechnology that may have adverse effects on biological diversity and human health'. Cartagena Protocol on Biosafety to the Convention on Biological Diversity, 11 September 2003, available at http://bch.cbd.int/protocol. Taken together, these two definitions encompass the spectre of measures, technologies, and procedures that are required to ensure occupational health and safety throughout the research process, as well as the safe handling of research results and products.

The importance of safe laboratory practice is recognised as an essential condition for conducting work in the area of synthetic biology. Following the first genetic engineering experiments in the 1970s which led scientists developing the technology to call for a research moratorium, heated debates on the future of work involving recombinant DNA (rDNA) resulted in an international consensus that research should continue but under stringent restrictions.[22] The pinnacle of these debates was the Asilomar Conference convened in 1975 which brought together some 140 participants including scientists, lawyers, journalists, and government officials. The recommendations of the conference largely informed the development of the official US guidelines for research involving rDNA molecules that were published a year later and, to date, are regularly updated.[23]

The issue of laboratory and environmental safety is explicitly acknowledged in the guide 'Doing Global Science: A Guide to Responsible Conduct in the Global Research Enterprise' that the Inter-Academy Partnership (IAP) published in 2016.[24] The guide defines the professional responsibilities of scientists and is intended as an essential tool for fostering the norms and principles of research integrity. Biosafety professional associations make a significant contribution to promoting safe work with biological materials through advocacy and capacity building, professional certification, and networking. For example, the International Federation of Biosafety Associations (IFBA) administers a Professional Certification Programme in different technical disciplines related to the management of biological risks and a Global Mentorship Programme that seeks to facilitate peer learning and experience sharing among practitioners. IFBA has also established a Biosafety Hero award, in order to celebrate the personal achievement of dedicated biosafety professionals and identify role models.[25]

---

[21] World Health Organisation, *Laboratory Biosafety Manual,* 4th ed., (Geneva: World Health Organisation, 2020), https://www.who.int/publications/i/item/9789240011311.

[22] P. Berg, 'Asilomar 1975: DNA Modification Secured', *Nature,* vol. 455 (2008), pp. 290–291, available at https://www.nature.com/articles/455290a.

[23] US National Institutes of Health, Biosafety and Recombinant DNA Policy, https://osp.od.nih.gov/biotechnology/biosafety-and-recombinant-dna-activities/.

[24] Inter-Academy Partnership, *Doing Global Science: A Guide to Responsible Conduct in the Global Research Enterprise* (Princeton NJ: Princeton University Press, 2016), available at https://www.interacademies.org/33345/Doing-Global-Science-A-Guide-to-Responsible-Conduct-in-the-Global-Research-Enterprise.

[25] For information about the activities of the International Federation of Biosafety Associations, see https://internationalbiosafety.org/.

Developing safe and good quality products is among the key priorities and responsibilities of any industry. In 2007, the Biotechnology Innovation Organisation (BIO) – the largest trade association representing private and public enterprises and academic institutions across the US and in over 30 other nations – launched the 'Excellence through Stewardship' (ETS) Programme, the first industry-coordinated effort to address product stewardship and quality management.[26] A year later, the programme evolved into a non-profit organisation which currently has over 50 members, including sector research institutions, technology providers, seed producers, and biotechnology associations from around the world.[27] The overriding goal of ETS is to enable enterprises to ensure effective compliance with the regulations that are applicable to their operations. To this end, ETS seeks to promote the universal adoption of quality management systems for the full life cycle of agricultural technology products through the articulation of relevant guiding principles and management practices; the development of training resources and programmes; and the administration of audit processes.

Contrary to common perceptions, the 'do-it-yourself' biology community has strived to internalise biosafety procedures and practices and ensure that these are tailored to the specific setting within which DIY biologists operate. As a result of a series of workshops and gatherings that brought together DIY practitioners from around the world, codes of conduct were developed in 2011 (Box 6.1).[28] These codes define a set of guiding principles by which practitioners agree to abide. Community laboratories have their own advisory boards comprising of technical experts who review project proposals and assist in addressing potential safety concerns. It is also possible for DIY biologists to seek advice and guidance from biosafety professionals via designated online portals.[29]

Similar to biosafety, the term 'biosecurity' has multiple definitions. The World Health Organisation considers biosecurity within the laboratory setting and defines laboratory biosecurity as the protection control and accountability for biological materials within laboratories, in order to prevent their unauthorised access, loss, theft, misuse, diversion, or intentional release.[30] More generally, biosecurity refers to the successful minimising of the risks that the biological sciences might be

---

[26] Biotechnology Innovation Organisation, 'BIO Launches the Excellence Through Stewardship Program Initiative Introduces Best Practices for Quality Management of Plant Biotechnology Products', *Press Release,* 25 July 2007, available at https://archive.bio.org/media/press-release/bio-launches-excellence-through-stewardship-program-initiative-introduces-best-p. See also Biotechnology Innovation Organisation, 'BIO Statement of BIO Ethical Principles', https://www.bio.org/articles/bio-statement-of-bio-ethical-principles.

[27] For information about the Excellence through Stewardship Organisation, see https://www.excellencethroughstewardship.org/.

[28] For information about the DIY biology codes of conduct, see https://diybio.org/codes/.

[29] T. Kuiken, 'Learn from DIY Biologists', *Nature,* vol. 531 (2016), available at https://www.nature.com/news/governance-learn-from-diy-biologists-1.19507; T. Landrain et al. 'Do-It-Yourself Biology: Challenges and Promises for an Open Science and Technology Movement', *Systems and Synthetic Biology,* vol. 7 (2013), pp. 115–126.

[30] World Health Organisation, *Biorisk Management: Laboratory Biosecurity Guidance* (Geneva: World Health Organisation, 2006), available at https://www.who.int/ihr/publications/WHO_CDS_EPR_2006_6/en/.

Box 6.1 DIYbio Codes of Ethics

European Congress: Draft DIYbio code of ethics	North American Congress: Draft DIYbio code of ethics
Transparency Emphasize transparency and the sharing of ideas, knowledge, data and results.	**Open Access** Promote citizen science and decentralized access to biotechnology.
Safety Adopt safe practices.	**Transparency** Emphasize transparency, the sharing of ideas, knowledge and data.
Open Access Promote citizen science and decentralized access to biotechnology.	**Education** Engage the public about biology, biotechnology and their possibilities.
Education Help educate the public about biotechnology, its benefits and implications.	**Safety** Adopt safe practices.
Modesty Know you don't know everything.	**Environment** Respect the environment.
Community Carefully listen to any concerns and questions and respond honestly.	**Peaceful Purposes** Biotechnology should only be used for peaceful purposes.
Peaceful Purposes Biotechnology must only be used for peaceful purposes.	**Tinkering** Tinkering with biology leads to insight; insight leads to innovation.
Respect Respect humans and all living systems.	
Responsibility Recognize the complexity and dynamics of living systems and our responsibility towards them.	
Accountability Remain accountable for your actions and for upholding this code.	

accidentally or deliberately misused in a way that causes harm to humans, animals, plants, or the environment.[31] This includes the risk of bioterrorism, bio-crimes, and development of biological weapons.

Biosecurity risks have attracted considerable attention over the past two decades, particularly in the light of rapid global diffusion of enabling capabilities with dual-use potential. The accidental discovery of a method for enhancing the virulence of the Mousepox virus, the artificial synthesis of the polio virus, the recreation of the

---

[31] S. Whitby et al. eds. *Preventing Biological Weapons: What You Can Do* (Bradford: University of Bradford, 2015), available at https://bradscholars.brad.ac.uk/handle/10454/7821.

Spanish Influenza virus, and the creation of a novel synthetic life form are among the early studies which have underscored the need for a careful assessment of the broader social, ethical, and legal implications of synthetic biology.[32] Two high-level reports published by the US National Research Council in 2004 and 2006, respectively have made recommendations in this regard. The Fink Committee report titled 'Biotechnology Research in an Age of Terrorism' defines seven types of experiments that require review by informed members of the scientific and medical community before they are undertaken or, if carried out, before they are published in full detail. These include experiments that:

1. Would demonstrate how to render a vaccine ineffective.
2. Would confer resistance to therapeutically useful antibiotics or antiviral agents.
3. Would enhance the virulence of a pathogen or render a non-pathogen virulent.
4. Would increase transmissibility of a pathogen.
5. Would alter the host range of a pathogen.
6. Would enable the evasion of diagnostic/detection modalities.
7. Would enable the weaponisation of a biological agent or toxin.[33]

The proposed criteria could serve as the backbone of an oversight system for minimising potential biosecurity concerns. However, as noted by the Committee, (1) the scope of the criteria is limited, since they address only microbial threats and (2) in the future, the proposed categories need to be expanded to cover a significantly wider range of potential threats.[34] The Lemon-Relman Committee report titled 'Globalisation, Biosecurity, and the Future of Life Sciences' has proposed a conceptual framework for assessing the potential for beneficial and disruptive

---

[32] See S. Whitby and M. Dando, 'Biosecurity Awareness-Raising and Education for Life Scientists: What Should Be Done Now?' in B. Rappert, ed. *Education and Ethics in the Life Sciences: Strengthening the Prohibition of Biological Weapons* (Canberra: ANU Press, 2010), pp. 179–196, available at https://press.anu.edu.au/publications/series/practical-ethics-public-policy/education-and-ethics-life-sciences. On examples of dual-use research, see M.Selgelid and L. Weir, 'The Mousepox Experience', *EMBO Reports,* vol. 11:1 (2010), pp. 18–24, available at https://www.embopress.org/doi/10.1038/embor.2009.270; E. Wimmer, 'The Test-Tube Synthesis of a Chemical Called Poliovirus: The Simple Synthesis of a Virus Has Far-Reaching Societal Implications', *EMBO Reports,* vol. 7: Spec No (2006), pp. S3–S9, available at https://www.embopress.org/doi/10.1038/sj.embor.7400728; J van Aken, 'Ethics of Reconstructing Spanish Flu: Is it Wise to Resurrect a Deadly Virus', *Heredity,* vol. 98 (2007), pp. 1–2, available at https://www.nature.com/articles/6800911; A. Katsnelson, 'Researchers Start Up Cell with Synthetic Genome', *Nature,* 20 May 2010, available athttps://www.nature.com/news/2010/100520/full/news.2010.253.html; 'Sizing up the "Synthetic Cell"', *Nature,* 20 May 2010, available at https://www.nature.com/news/2010/100520/full/news.2010.255.html.

[33] National Research Council, *Biotechnology Research in an Age of Terrorism* (Washington DC: National Academies Press, 2004), available at https://www.nap.edu/catalog/10827/biotechnology-research-in-an-age-of-terrorism.

[34] National Research Council, *Biotechnology Research in an Age of Terrorism* (Washington DC: National Academies Press, 2004), available at https://www.nap.edu/catalog/10827/biotechnology-research-in-an-age-of-terrorism.

applications of the novel life science advances. [35] The Committee has developed a system of classification comprising of four thematic groupings, namely:

1. Technologies that seek to acquire novel biological or molecular diversity;
2. Technologies that seek to generate novel but pre-determined and specific biological or molecular entities through directed design;
3. Technologies that seek to understand and manipulate biological systems in a more comprehensive and effective manner;
4. Technologies that seek to enhance production, delivery, and "packaging" of biologically active materials.

The report has recommended that a broader perspective on the 'threat spectrum' is adopted by focusing on trends in life science advances that can facilitate hostile misuse.[36]

Unlike biosafety considerations, by and large, biosecurity risks may not be immediately evident to life science stakeholders. A case in point in this regard is the multifaceted international controversy that spurred as a result of the creation of a mammalian-transmissible H5N1 virus in 2011.[37] The two studies conducted independently in the Netherlands and the USA met several of the criteria for experiments of concern as defined by the Fink Committee. In 2005, the Inter-Academy Panel published a Statement on Biosecurity which acknowledged the special responsibility of scientists regarding problems of dual use and the misuse of science and technology and the duty to be aware and foresee the possible consequences of their own activities.[38] One of the lead scientists of the Dutch research team had participated in the focus group established to support the development of the Code of Conduct on Biosecurity that the Royal Netherlands Academy of Arts and Sciences had adopted 4 years earlier (Box 6.2).[39] Nevertheless, biosecurity issues were only considered after the editorial boards of *Science* and *Nature* decided to defer the publication of the manuscripts and the papers were subject to additional review.[40] The publication of the two studies was preceded by a protracted global debate on

---

[35] National Research Council, *Globalisation, Biosecurity, and the Future of Life Sciences* (Washington DC: National Academies Press, 2006), available at https://www.nap.edu/catalog/11567/globalization-biosecurity-and-the-future-of-the-life-sciences.

[36] National Research Council, *Globalisation, Biosecurity, and the Future of Life Sciences* (Washington DC: National Academies Press, 2006), available at https://www.nap.edu/catalog/11567/globalization-biosecurity-and-the-future-of-the-life-sciences.

[37] On the H5N1 controversy, see *Nature* Special Collection, available at https://www.nature.com/collections/wntqfnjrxb.

[38] Inter-Academy Panel, *IAP Statement on Biosecurity,* 2005, available at https://www.interacademies.org/13912/IAP-Statement-on-Biosecurity.

[39] Royal Netherlands Academy of Arts and Sciences (KNAW), *A Code of Conduct for Biosecurity* (Amsterdam: Royal Netherlands Academy of Arts and Sciences, 2008), available at https://www.knaw.nl/en/news/publications/a-code-of-conduct-for-biosecurity.

[40] For a review of editorial policies regarding the publication of dual-use research of concern, see D. Patrone et al. 'Biosecurity and the Review and Publication of Dual-Use Research of Concern', *Biosecurity and Bioterrorism,* vol. 10:3 (2012), pp. 290–298, available at https://www.liebertpub.

how the risks and benefits of life science research should be balanced has demonstrated that, by and large, biosecurity issues tend to be considered mainly within the context of laboratory practice, whereby priority is given to the physical security of biological materials and information, including through access control and vetting of research personnel. The debate has further shown that concerns of dual use and science misuse are rarely considered and addressed at the different stages of research process.

In its 2016 publication, 'Doing Global Science: A Guide to Responsible Conduct in the Global Research Enterprise', the IAP has noted that preventing the misuse of life science research is likely to challenge researchers and the broader research enterprise in future which is why researchers need to participate in discussions about the possible consequences of their work, including harmful consequences, when planning research projects.[41] More recently, the World Health Organisation has sought to provide additional guidance on the governance of dual-use research of concern (DURC) in the life sciences – "research that, based on current understanding, has the potential to provide knowledge, information, products or technologies that could be directly misapplied to create a significant threat with potential consequences to public health and safety, agricultural species and other plants, animals, and the environment".[42] According to WHO, the recommended approach for DURC management is "laboratory and medical-scientific self-governance" underpinned by regulatory oversight and "an enhanced culture of trust, personal responsibility, accountability and transparency in laboratories, a culture which comes from strong leadership and a commitment to championing ethics in the workplace".[43]

---

com/doi/10.1089/bsp.2012.0011. Both the Dutch and US research papers were eventually published in 2012, see S. Herfst et al. 'Airborne Transmission of Influenza A/H5N1 Virus Between Ferrets', *Science,* vol. 336:6088 (2012), pp. 1534–1541, DOI: https://doi.org/10.1126/science.1213362; M. Imai et al. 'Experimental Adaptation of an Influenza H5 HA Confers Respiratory Droplet Transmission to a Reassortant H5 HA/H1N1 Virus in Ferrets', *Nature,* vol. 486 (2012), pp. 420–428, https://www.nature.com/articles/nature10831.

[41] Inter-Academy Partnership, *Doing Global Science: A Guide to Responsible Conduct in the Global Research Enterprise* (Princeton NJ: Princeton University Press, 2016), available at https://www.interacademies.org/33345/Doing-Global-Science-A-Guide-to-Responsible-Conduct-in-the-Global-Research-Enterprise.

[42] World Health Organisation, *Laboratory Biosafety Manual,* 4th ed. (Geneva: World Health Organisation, 2020), https://www.who.int/publications/i/item/9789240011311.

[43] World Health Organisation, *WHO Guidance on Implementing Regulatory Requirements for Biosafety and Biosecurity in Biomedical Laboratories – A Stepwise Approach* (Geneva: World Health Organisation, 2020), https://www.who.int/publications/i/item/who-guidance-on-implementing-regulatory-requirements-for-biosafety-and-biosecurity-in-biomedical-laboratories%2D%2Da-stepwise-approach.

Box 6.2 The Dutch Code of Conduct for Biosecurity [Emphases Added]
In 2007, the Royal Netherlands Academy of Arts and Sciences (KNAW) adopted a 'A Code of Conduct for Biosecurity'.[44] The Code aims to **prevent** life sciences research or its application from **directly or indirectly** contributing to the development, production or stockpiling of biological weapons, as described in the **BTWC**, or to any other **misuse** of biological agents and toxins. It targets different groups of stakeholders and defines six basic principles of biosecurity including:

- Raising awareness.
- Research and publication policy.
- Accountability and oversight.
- Internal and external communication.
- Accessibility.
- Shipment and transport.[45]

Following the H5N1 controversy, in 2013, the Royal Netherlands Academy of Arts and Sciences published a report titled 'Improving Biosecurity: Assessment of Dual-Use Research' which underscored that **'the primary responsibility for dealing with potential dual-use risks of life science research lies with the researchers and parties in the knowledge chain'**.[46]

The report outlined a biosecurity assessment framework noting that 'when determining whether a study should be regarded as dual use from the perspective of biosecurity, both the **biological** and the **contextual** factors must be considered. […]. The question then is not only **whether** a research project is dual use within the context of biosecurity, but in particular **what consequences** this should have.'

The report further recommended the establishment of a Biosecurity Advisory Committee in the Life Sciences. The proposed Advisory Committee would fulfil both **case-specific tasks**, such as advising on specific research proposals, reviewing reports by whistle-blowers about projects and

(continued)

[44] Royal Netherlands Academy of Arts and Sciences (KNAW), *A Code of Conduct for Biosecurity* (Amsterdam: Royal Netherlands Academy of Arts and Sciences, 2008), available at https://www.knaw.nl/en/news/publications/a-code-of-conduct-for-biosecurity.

[45] Royal Netherlands Academy of Arts and Sciences (KNAW), *A Code of Conduct for Biosecurity* (Amsterdam: Royal Netherlands Academy of Arts and Sciences, 2008), available at https://www.knaw.nl/en/news/publications/a-code-of-conduct-for-biosecurity.

[46] Royal Netherlands Academy of Arts and Sciences, Biosecurity Committee, *Improving Biosecurity: Assessment of Dual Use Research* (Amsterdam: Royal Netherlands Academy of Arts and Sciences, 2013), available at https://www.knaw.nl/en/news/publications/improving-biosecurity.

Box 6.2 (continued)

researchers, and reporting, as well as **system-based tasks**, such as keeping track of scientific, technological and policy-related trends and developments, maintaining contacts with research institutions, international networking, facilitating public engagement, communication, and accountability.

Finally, the report highlighted that the 'Code of Conduct for Biosecurity **should be an ongoing topic of interest in education and researcher training** and for research team heads and funding bodies'.[47]

6.4 Enhancing Stakeholder Interaction in the Field of Biosafety and Biosecurity

The professional diversity within the synthetic biology community presupposes a multitude of professional cultures, each characterised by its own system of values, shared meanings, established practices, and routines. These cultures are constantly in flux and being conditioned by the larger national cultures within which they exist and operate. Each professional culture is a manifestation of the prevalent priorities and objectives that different stakeholders set and pursue. The ways in which the concepts of risks and benefits are framed by different stakeholders inevitably vary, not least because these concepts are expressions of the dominant common understandings and interests that each professional group considers important. A robust biosafety and biosecurity culture entails the existence of mechanisms, practices, procedures, and attitudes which ensure that risks and concerns are raised, tackled, and effectively managed throughout the full research and innovation cycle.[48] Active interaction among stakeholders is crucial for finding a common ground for constructive dialogue and identifying viable avenues for reconciling competing interests among different professional cultures. Cooperation is key in order to develop and implement adequate and sustainable approaches for risk mitigation which do not hinder research and innovation.

Education and training are key elements of the process of sensitising prospective and practising scientists to the values of research integrity, responsible conduct, and professionalism. Science classes are meant to encourage curiosity and desire to learn and aspire. They also provide an opportunity to foster an understanding of the social responsibility of scientists to be aware of the broader implications of their work and carry out an informed assessment of the risks and benefits involved (Box 6.3).

---

[47] Royal Netherlands Academy of Arts and Sciences, Biosecurity Committee, *Improving Biosecurity: Assessment of Dual Use Research* (Amsterdam: Royal Netherlands Academy of Arts and Sciences, 2013), available at https://www.knaw.nl/en/news/publications/improving-biosecurity.

[48] National Academies of Sciences, Engineering, and Medicine, *Governance of Dual-Use Research in the Life Sciences: Advancing Global Consensus on Research Oversight* (Washington DC: National Academies Press, 2018), available at https://www.nap.edu/catalog/25154/governance-of-dual-use-research-in-the-life-sciences-advancing.

Box 6.3 International Recognition of the Need for Responsible Science Education
NRC (US), *Biotechnology Research in an Age of Terrorism,* 2004:[49]

Recommendation 1: Educating the Scientific Community

The Committee has recommended that 'national and international professional societies and related organizations and institutions create programs to educate scientists about the nature of the dual use dilemma in biotechnology and their responsibilities to mitigate its risks.'

NRC (US), *Globalisation, Biosecurity, and the Future of the Life Sciences,* 2006:[50]

Recommendation 4: The committee recommends the adoption and promotion of a common culture of awareness and a shared sense of responsibility within the global community of life scientists.

4a. 'Recognize the value of formal international treaties and conventions, including the 1972 Biological and Toxin Weapons Convention (BWC) and the 1993 Chemical Weapons Convention (CWC).
4b. Develop explicit national and international codes of ethics and conduct for life scientists.
4c. Support programs promoting beneficial uses of technology in developing countries.
4d. Establish globally distributed, decentralized, and adaptive mechanisms with the capacity for surveillance and intervention in the event of malevolent applications of tools and technologies derived from the life sciences.'

German Ethics Council, *Biosecurity – Freedom and Responsibility of Research,* 2014:[51]

Recommendation 1: Raising the level of awareness for questions of biosecurity in the scientific community

'In view of the potential for misuse of dual use research in the life sciences, there is a need to increase the degree of awareness amongst members of the

(continued)

[49] National Research Council, *Biotechnology Research in an Age of Terrorism* (Washington DC: National Academies Press, 2004), available at https://www.nap.edu/catalog/10827/biotechnology-research-in-an-age-ofterrorism.

[50] National Research Council, *Globalisation, Biosecurity, and the Future of Life Sciences* (Washington DC: National Academies Press, 2006), available at https://www.nap.edu/catalog/11567/globalization-biosecurity-andthe-future-of-the-life-sciences.

[51] German Ethics Council, *Biosecurity – Freedom and Responsibility of Research,* German Ethics Council, 2014, available at https://www.ethikrat.org/fileadmin/Publikationen/Stellungnahmen/englisch/opinion-biosecurity.pdf.

Box 6.3 (continued)

scientific community for these issues and to promote an underlying culture of responsibility.'

Swiss Academies of Arts and Sciences, *Misuse Potential and Biosecurity in Life Sciences Research*, 2017:[52]

'**Education** and **training** in **biosecurity** are among the most effective strategies to anticipate and prevent misuse of life science research. [...] Other important measures to prevent misuse include fostering **responsible** research practices and **scientific integrity** more generally and cultivating an atmosphere of **trust** at research institutions and in research groups.'

Continued professional development training allows practising researchers to keep up to date with relevant policy and legislative developments and ensure that institutional procedures and practices are aligned with national regulations. It is important that biosafety and biosecurity issues are given equal attention during education and training and that the complementary role of biosafety and biosecurity in the governance of science and technology is elucidated. A case in point is the Professional Certification Programme of the International Federation of Biosafety Association which features Biosecurity as a technical discipline. The Biosecurity certification exam covers six topic areas:

(1) Biosecurity Conventions, Guidelines and Standards;
(2) Biosecurity Risk Assessment and Programme Management;
(3) Physical Biosecurity Measures;
(4) Pathogen Accountability;
(5) Personnel Reliability;
(6) Dual-use and Bioethics.[53]

Those willing to sit the exam need to hold a valid certification in Biorisk Management which covers basic laboratory biosafety concepts, among other things. The Biosecurity Professional Certification aims to promote biosecurity learning and competence among practising researchers, so that they can subsequently apply the acquired skills and knowledge on their workplace, for example, by helping introduce biosecurity concepts into the existing institutional oversight policies and staff development training schemes.

When implementing biosafety and biosecurity education and awareness-raising programmes, attention needs to be given both to the content and mode of its

---

[52] Swiss Academies of Arts and Sciences, *Misuse Potential and Biosecurity in Life Sciences Research: A Discussion Basis for Scientists on How to Address the Dual Use Dilemma of Biological Research* (Swiss Academies Report: 2017), available at https://naturalsciences.ch/organisations/geneticresearch/topics/biosecurity

[53] For information about the IFBA Professional Certification in Biosecurity, 2020, see https://internationalbiosafety.org/wp-content/uploads/2019/02/3.2-Professional-Certification-in-Biosecurity-Exam-Content-English.pdf. See also R. Moritz et al. 'Promoting Biosecurity by Professionalizing Biosecurity', *Science*, vol. 367:6480 (2020), pp. 856–858, https://science.sciencemag.org/content/367/6480/856https://science.sciencemag.org/content/367/6480/856.

delivery. Teaching and training methods need to be carefully selected, in order to maximise learning impact and facilitate the application of relevant knowledge to everyday science practice.[54] Active learning techniques such as simulations and scenario-based exercises encourage critical reflection and self-assessment, and contribute to an enhanced understanding of biosafety and biosecurity risks. This in turn enables stakeholders to be proactive in the process of risk governance and develop a sense of ownership. An in-depth shared understanding of the risks posed by advances in synthetic biology among stakeholders is essential, in order to ensure consistency and coherence across the implemented mechanisms and approaches.

Examples of initiatives that seek to promote responsible innovation in the field of synthetic biology include the Engineering Biology Research Consortium (EBRC) and the International Genetically Engineered Machine (iGEM) Competition. EBRC is a non-profit, public-private partnership dedicated to advancing engineering biology.[55] EBRC administers a programme on improving security considerations that is designed to facilitate education and dialogue on security issues among stakeholders through workshops, awareness-raising, and development of training material. Launched in 2004, the iGEM Competition is an annual event that brings together interdisciplinary teams of university and high school students, DIY biologists, and more from around the world and provides them with the opportunity to push the boundaries of synthetic biology by tackling everyday social and environmental challenges.[56] iGEM has a dedicated biosafety and biosecurity program which operates throughout the life cycle of projects – from inception to future applications – allowing risks and concerns to be identified, flagged up, and addressed in a timely manner (Box 6.4).[57] iGEM participants also have at their disposal the 'iGEMers Guide to the Future' which is an online resource designed to provide iGEM participants with a space, process, and tools for facilitating project development and responsible design and innovation.[58] The Guide has been developed as a result of a EU-funded collaborative initiative titled 'Synthetic Biology – Engaging with New and Emerging Science and Technology in Responsible Governance of the Science and Society Relationship' (SYNENERGENE), designed to establish an open dialogue among stakeholders on the potential benefits and risks of synthetic biology.[59]

---

[54] T. Novossiolova et al. 'Altering an Appreciative System: Lessons from Incorporating Dual-Use Concerns into the Responsible Science Education of Biotechnologists', *Futures*, vol. 108 (2019), pp. 53–60, available at https://www.sciencedirect.com/science/article/abs/pii/S001632871830466X.

[55] For information about the Engineering Biology Research Consortium, 2020, see https://ebrc.org/.

[56] For information about the International Genetically Engineered Machine Competition, 2020, see https://igem.org/Main_Page.

[57] P. Millet et al. 'Developing a Comprehensive, Adaptive, and International Biosafety and Biosecurity Program for Advanced Biotechnology: The iGEM Experience', *Applied Biosafety*, vol. 24:2 (2019), pp. 64–71, available at https://journals.sagepub.com/doi/full/10.1177/1535676019838075.

[58] The 'iGEMers Guide to the Future' is available at https://live.flatland.agency/12290417/rathenau-igem/.

[59] The SYNENERGENE project was carried out between July 2013 and June 2017 as part of the FP 7 funding scheme of the European Commission. Further information about the initiative is available at https://www.synenergene.eu/index.html.

Box 6.4 iGEM Biosafety and Biosecurity Programme[60]

iGEM's biosafety and biosecurity programme is forward-leaning, in that it addresses both traditional (pathogen-based) and emerging risks both in terms of new technologies and new risks. It is integrated into the technical work of the competition – with clearly described roles and responsibilities for all members of the community. The program makes use of both incentives (such as through a Safety and Security Award for excellence and human practices components of its medals) and penalties for noncompliance (up to and including disqualification).

As all biological lab work, even simple experiments, carries some risk, teams must follow a set of safety and security rules:

- Teams must be in full compliance with iGEM's safety and security policies.
- Teams must use the competition's forms to provide information on any risks from their project and steps taken to manage them.
- The Safety and Security Committee must have approved (a) check-in forms before a team uses parts and organisms not on the white list and (b) animal use forms before teams use vertebrates and some invertebrates.
- Instructors must sign off relevant forms.
- All deadlines for providing safety and security information must be met.
- Teams must follow all relevant international, regional, national, local, or institutional laws, rules, regulations, or policies, including national or institutional biosafety and biosecurity rules. If conducting any experiment with human subjects (including noninvasive experiments, such as surveys), teams must comply with all rules governing experiments with human subjects.
- Teams must work in the biosafety level appropriate for their project.
- Teams cannot conduct work with risk group 3 or 4 organisms, parts from a risk group 4 organism, or work in a safety level 3 or 4 laboratory.
- Teams must follow iGEM shipment requirements when submitting samples.
- Teams cannot release or deploy their project outside of the laboratory (including putting them in people) at any time during the competition or at the Giant Jamboree.

Similar to iGEM, DIY biology community labs have been recognised as potential catalysts for promoting responsible innovation.[61] 'Patient-led research' or 'citizen-driven biomedical research' is a new form of research where citizens and patients are the primary *producers* and *mobilizers* or *instigators* of knowledge

[60] This text box is based on Piers Millet et al. 'Developing a Comprehensive, Adaptive, and International Biosafety and Biosecurity Program for Advanced Biotechnology: The iGEM Experience', *Applied Biosafety,* vol. 24:2 (2019), pp. 1–8.

[61] E. Pauwels and S. Denton, *The Rise of the Bio-Citizen,* Wilson Center, January 2018, available at https://www.wilsoncenter.org/article/the-rise-the-new-bio-citizen.

pursuing a range of activities from analyses of genomic data for diagnosing rare diseases, identification of potential therapeutic drugs, organization and crowdfunding of clinical trials' cohorts, and even self-surveillance or self-experimentation.[62] Collectively the DIYbio community have adapted and adopted biosafety standards to meet their needs as well as worked with ABSA International to develop a biosafety boot-camp training program in order to promote mentorship regarding the risks and benefits of emerging technologies.[63]

6.5 Conclusion

This chapter has sought to examine the role that the synthetic biology community can play in addressing the security implications of their work and thus contribute to the efforts to ensure that the life sciences are used only for peaceful purposes. The interconnectedness of the following three points is of particular importance in this regard:

- Identifying, assessing, and mitigating biosafety and biosecurity risks related to emerging life science advances (e.g. synthetic biology) requires the active engagement of all science stakeholders, including professional and amateur science practitioners.
- There is a need for institutionalised early and recurring training in responsible conduct of research, biosafety, and biosecurity for prospective and practising scientists and engineers, in order to foster a shared understanding of the potential risks and how they can be addressed. Equally, it is important that DIY biology communities internalise the requirements for biosafety and biosecurity awareness and practices.
- Stakeholder interaction, experience sharing, and collaboration among the different professional and non-professional communities engaged in synthetic biology is vital to strengthening the concepts of stewardship, responsibility, and accountability, in order to safeguard research and innovation against accidental or deliberate misuse.

---

[62] E. Pauwels and S. Denton, *The Rise of the Bio-Citizen,* Wilson Center, January 2018, available at https://www.wilsoncenter.org/article/the-rise-the-new-bio-citizen.

[63] Lim, Y. B., *Checking Ourselves Before Wrecking Ourselves: Co-Evolving Innovation and Safety in the DIYBio Community*, BUGSS, September 2019, available at https://bugssonline.org/community/diybio-biosafety/. See also L. Sundaram, 'Biosafety in DIY-Bio Laboratories: From Hype to Policy', *EMBO Reports*, e52506 (2021), available at https://www.embopress.org/doi/abs/10.15252/embr.202152506

Acknowledgement Dr Lorna Miller (ret.), Defence Science and Technology Laboratory, UK and Prof Jose Manuel Palma-Oliveira, University of Lisbon, Portugal provided invaluable insights and input during the preliminary considerations that underpinned the development of this chapter.

Disclaimer The views expressed here are those of the authors alone. They may not represent the views of their affiliated organisations and should not be taken as an official statement or position of the affiliated organisations.

7

Covid-19 Pandemic and Cyberbiosecurity Threats

Aaron Adler, Jake Beal, Mary Lancaster and Daniel Wyschogrod

7.1 Introduction

Cyberbiosecurity, the aspect of biosecurity involving the digital representation of biological data, had already been emerging as a matter of public concern even prior to the onset of the COVID-19 pandemic. Key issues of concern include, among others, the privacy of patient data, the security of public health databases, the integrity of diagnostic test data, the integrity of public biological databases, the security implications of automated laboratory systems and the security of proprietary biological engineering advances.

With the onset of the COVID-19 pandemic, and the importance of digital resources in combatting it, concern about the potential for cyber attacks by state-based or non-state actors has been elevated. To illuminate the challenges, we focus on the cyber vulnerabilities that need to be addressed in public health activities such as disease surveillance and outbreak management. In particular, we examine cyber issues raised by the accelerated pace of development for COVID mitigations, treatments, and vaccines.

Figure 1 illustrates a simplified view of key components and their interactions in this area, as well as vulnerable points where informational attacks can result in significant biosecurity consequences. In particular, the challenges that we consider here are:

1. Privacy of contact tracing data – Contact tracing has been used in one form or another to contain epidemics for centuries. With the widespread adoption of

A. Adler · J. Beal · D. Wyschogrod (✉)
Raytheon BBN, Cambridge, MA, USA
e-mail: dan.wyschogrod@raytheon.com

M. Lancaster
Pacific Northwest National Laboratory (PNNL), Richland, WA, USA

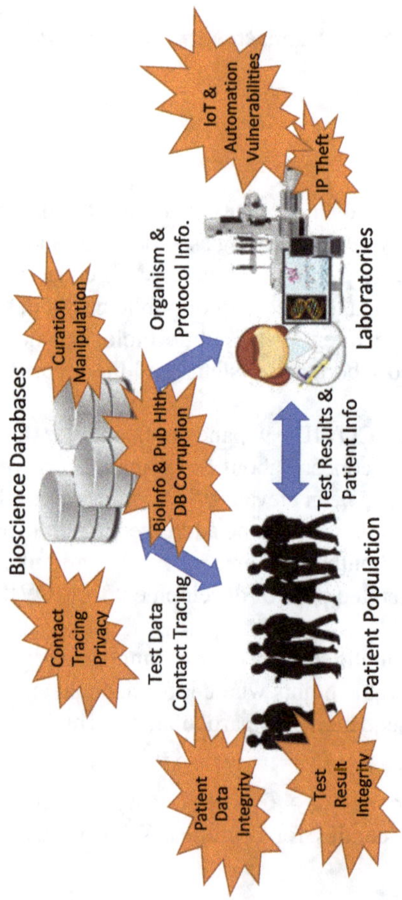

Fig. 1 Key components, interactions, and cyberbiosecurity threats in the surveillance and management of infectious diseases such as COVID-19

smart phones, the potential for automated contact tracing holds significant promise. How can this be done in a manner that protects patient privacy? How can smartphone data be integrated with manual contact tracing? What are the privacy, security and efficacy tradeoffs? What are the implications of patient privacy concerns for the collection of public health data?

2. Integrity of public health and disease surveillance data – As COVID-19 has shown, disease surveillance data is critical both for scientists and policy makers. These data include but are not limited to case counts, diagnostic test results, and general trend information. Various governments, agencies, or other malicious actors might want to manipulate such information to artificially inflate or suppress data. What safeguards against such manipulation can be provided?
3. Data integrity and result validation of self-administered testing – Self-administered tests can provide fast, actionable health information. For COVID-19, a number of at home tests are being proposed, some of which would allow users to receive immediate results, similar to pregnancy tests. Self-administration, however, also allows many more opportunities for data corruption or exposure. How can such results be shared for aggregation into public health statistics and use in contact tracing in such a way that their results can be validated by health care professionals and individual privacy also be preserved?
4. Integrity of public bioinformatic databases – Both researchers and medical personnel rely on public sequence and sample data resources such as those maintained by NCBI. Frequently, mistakes are made in labeling that can cause difficulties. Currently, most such errors appear to be inadvertent rather than malicious, but such data could also be deliberately manipulated to confuse bioinformatic investigations. How can the integrity of public data be maintained and attempts at manipulation detected?
5. Defending against cyberattacks on laboratory automation – Laboratory throughput is increasingly being accelerated through automation involving robotics, laboratory information management systems (LIMS), and network-enabled devices that fall under the general category of Internet of Things (IoT). These systems are often connected to the internet (e.g., for software updates or remote monitoring and control), providing an attack surface by which they may be compromised. Such devices can be used as entry points into laboratory networks or manipulated for their biological effects (e.g., destroying stored specimens by changing temperature settings on a freezer). How can these devices be protected and how can they be prevented from becoming points of entry into critical laboratory computer networks?
6. Protection of intellectual property – Theft of intellectual property by both state and non-state actors is a longstanding problem. The race for COVID-19 vaccinations and treatments has amplified this illicit activity, motivated both by a desire for direct monetary gain and by nations' need to protect their populations and restore their economies. What safeguards need to be provided and how can malicious parties be identified?

In the subsequent sections, we will expand on each of these threats in turn, followed by a summary of their implications.

7.2 Privacy of Contact Tracing Data

Contact tracing is the process of identifying and monitoring persons who have been in contact with an infected person or persons. It has been used in one form or another for centuries.[1] More recently, it has been used effectively in the control of tuberculosis, Severe Acute Respiratory Syndrome (SARS), and Middle East Respiratory Syndrome (MERS).[2] Manual contact tracing has limitations in the number of persons that can be identified and interviewed in a timely manner. With about 8000 SARS infections and 800 deaths[3] and about 2500 instances of MERS and 858 deaths,[4] manual contact tracing proved sufficient. In the case of COVID-19, with 4.3 million confirmed cases and about 300,000 deaths worldwide at the time of this writing, complete and timely manual contact tracing may not be possible in many local jurisdictions.

Early in the COVID-19 pandemic, a number of countries leveraged smart phones to help automate contact tracing. Various types of relevant information are available on a smart phone. GPS and location services, which can add information about nearby WiFi hotspots, can be used with contact tracing apps but raises privacy concerns. In China, people are sent QR codes on their phones indicating their level of risk for COVID-19 and access to public transportation or public areas such as shopping malls is determined by the QR code granted to an individual.[5] These codes are based on self-reported information as well as possibly location services information (though the Chinese government has not been forthcoming on the data used to produce these codes). South Korea does not use such QR health codes, but publicizes details concerning individuals who have tested positive including the person's age range, gender, and places they recently visited. QR codes can also be used to register visitors to businesses and users of public transportation.

---

[1] S. Cohen, M. O'Brian, *The Conversation*, 'Contact tracing: how physicians used it 500 years ago to control the bubonic plague', https://theconversation.com/contact-tracing-how-physicians-used-it-500-years-ago-to-control-the-bubonic-plague-139248, June, 2020 (retrieved August 2020).

[2] K.O. Kwok, A. Tang, V.W.I. Wei, W. H. Park, E.K. Yeoh, and S. Riley, "Epidemic Models of Contact Tracing: Systematic Review of Transmission Studies of Severe Acute Respiratory Syndrome and Middle East Respiratory Syndrome", Comput Struct Biotechnol J., 2019; 17;186–194

[3] CDC, "Fact Sheet: Basic Information about SARS", https://www.cdc.gov/sars/about/fs-SARS.pdf, retrieved August 2020.

[4] WHO, "Middle East respiratory syndrome coronavirus (MERS-CoV)", https://www.who.int/emergencies/mers-cov/en/, retrieved August 2020.

[5] BBC, "China launches coronavirus 'close contact detector' app", https://www.bbc.com/news/technology-51439401, February 2020 (retrieved August 2020).

Another approach that is believed to be more privacy preserving and more secure in a number of respects involves the use of Bluetooth rather than GPS or location services. Singapore has released an app called TraceTogether.[6] TraceTogether attempts to minimize the amount of personal information it gathers, but it does collect the cell phone numbers of users on a voluntary basis.[7]

The ability to use Bluetooth and maintain a high level of privacy has been greatly assisted by the cooperation of Google and Apple in inserting new capabilities in both iOS and Android at the operating system level.[8] The Apple/Google protocol is based on privacy ideas emerging from the MIT-led PACT project[9] and the European DP-3T[10] project.

The goal of the Apple/Google application protocol interface (API) is to provide a set of functions and procedures in the operating system that can be used by state or local authorities and software developers to develop user-level contact tracing apps. The two foundations of this methodology are:

1. Extensions of the Bluetooth protocol to determine "too close for too long"
2. A distributed architecture such that notifications of proximity to a confirmed case of COVID-19 are sent only to the user of a phone and no other parties.

Algorithms to determine "too close for too long," however, are still under development. They have both a physical and biological component. The physical aspect involves the inference of distance between infectious and susceptible individuals from the observed information. The new interface provided by Apple/Google will give the app developer information about Received Signal Strength Indication (RSSI) for each transmission from a nearby (typically tens of meters) source.[11] The RSSI falls off with distance so it can be used to infer distance between phones but also falls off with attenuation due to phones being in pockets or handbags and intervening obstacles (e.g. walls or shelving), making the translation from RSSI to distance complex. The biological issue is how much exposure to an infected person at what distance indicates a high risk of infection. Tuning the criterion for "too close for too long" clearly will affect both the false positive and false negative rates.

The second foundation of this methodology is that only a user of the app is informed of a possible exposure, but no one else. The goal here is to achieve

---

[6] Singapore Government, "TraceTogether home page", https://www.tracetogether.gov.sg, retrieved August 2020.

[7] Singapore Government, "TraceTogether Privacy Safeguards", https://www.tracetogether.gov.sg/common/privacystatement, retrieved August 2020.

[8] Apple, Inc., "Privacy-Preserving Contact Tracing",https://covid19-static.cdn-apple.com/applications/covid19/current/static/contact-tracing/pdf/ExposureNotification-CryptographySpecification v1.2.pdf, retrieved August 2020.

[9] PACT, "PACT: Private Automated Contact Tracing", https://pact.mit.edu, retrieved August 2020.

[10] DP3T, "DP3T – Decentralized Privacy-Preserving Proximity Tracing" https://github.com/DP-3T/documents, retrieved August 2020.

[11] Bluetooth SIG, "Proximity and RSSI", https://www.bluetooth.com/blog/proximity-and-rssi/, September 2015 (retrieved, August 2020).

maximal privacy. The mechanism works as follows. An individual phone creates a seed at a particular time period, say each hour. That seed is used to generate changing values in each "chirp" emitted by the Bluetooth interface. Neighboring phones detect these chirps and record them along with timestamps. If an individual tests positive and they consent, their phone is accessed and the list of seeds over the infection time period are uploaded to a central database. The central database downloads the seeds with time stamps of all infected individuals to all users of the app. The user's phone then generates the seeds of infected persons to generate chirp values, which they check against the user's list of received chirps to see if the user has been exposed to any COVID-19 confirmed case. Thus, the recipient only knows that they have been potentially exposed to confirmed case of COVID-19. They do not know the identity of the person they were exposed to, nor does anyone else know that the user might be infected. The exposed app user is encouraged to seek diagnostic testing and to self-quarantine, but this is voluntary.

An important item to note is that while private information is withheld from unauthorized malicious or just curious agents, it is also withheld from health professionals and public health authorities, including human contact tracers. This information would undoubtedly be useful in determining with whom an infected individual came in contact, many of whom they may have forgotten or not noticed. While individuals who are notified about contact with infected individuals may be encouraged to contact health authorities, it would be voluntary and because of the anonymity protections, much of the work tracing back to previous contacts and forward to successive contacts would have to be repeated by the human contact tracers. Following the chain of individuals who are farther and farther removed from the diagnosed individual would be particularly useful for superspreader events where rapid identification and quarantine of all those exposed in the first several generations is critical. Identifying and isolating individuals with asymptomatic infections is also important. All of this information would have to be re-discovered by the human contact tracer.

A possible solution might be the voluntary submission of information to public health authorities by individuals who have gotten a match on their phone, perhaps through the app itself. The issue then is that more and more potentially private information is entered into the central cloud database.

These are all issues that are under active discussion. Since the Apple/Google interface is at the operating system and API level, however, different countries and regions will be able to choose to make different privacy decisions.

Such issues of privacy versus importance of data collection in emergency situations will apply to future post-COVID situations as well. Depending on the success of automated contact tracing in assisting in opening up commerce and day-to-day life, such apps, and perhaps their extension to wearables, may become more commonplace. Clearly, this is an issue where epidemiologists, infectious disease specialists, privacy and security experts, and medical ethicists must collaborate to identify and address risks and vulnerabilities.

7.3 Protecting Public Health and Disease Surveillance Data

As COVID-19 has shown, disease surveillance data are critical both for scientists and policy makers as well as the general public. These data include but are not limited to case counts, diagnostic test results, and general trend information. Various nation state or other malicious actors might want to either suppress or artificially inflate data.

As we have seen in the COVID-19 pandemic and the West Africa Ebola outbreak,[12] delayed response to outbreak events can result in larger impacts. If false negative diagnostic tests are returned and surveillance data are altered to keep case counts below epidemic thresholds, outbreak control measures may not be implemented until much later, when the outbreak is much larger.

Alternately, creation of disease cases in a surveillance system may create the appearance of an outbreak and result in mobilization of resources to investigate and mitigate an outbreak that does not exist. For example, for some livestock diseases, control measures include depopulation of the affected farms. Failure to properly confirm the presence of an outbreak before control measures are implemented may be catastrophic.

Furthermore, for many livestock and agricultural diseases, trade restrictions may be invoked to prevent the spread of disease across borders. The false creation of an outbreak in surveillance data may result in significant trade losses until the apparent outbreak can be invalidated. Considerable resources may be expended in verifying to trade partners and international organizations that an apparent outbreak was not real and that the animals in a herd or in a geographic area are not infected.

7.4 Integrity and Validation of Self-Administered Testing

Testing for SARS-CoV-2 provides a test case for the use of home diagnostic tests. Some routine tests, e.g., pregnancy tests or glucose tests, have long been available for home use. Some SARS-CoV-2 tests also allow for in-home collection, but specimens must be mailed to a laboratory that processes the results. This is similar to some other at-home testing systems, e.g., via Everlywell.[13] With SARS-CoV-2 testing, there is a public health interest in tracking test results, and integrity and validation are important.

There are four broad categories for handling of at-home testing:

1. Self tests with unreported results;
2. Self tests shared and interpreted via a telehealth appointment;

---

[12] M. Jeremiah Matson, Daniel S. Chertow, and Vincent J. Munster,"Delayed recognition of Ebola virus disease is associated with longer and larger outbreaks," Emerg Microbes Infect. 2020; 9(1): 291–301.

[13] Everywell, "Everywell home page", https://www.everlywell.com, retrieved August 2020.

3. Self tests interpreted via a cell phone mobile application;
4. Self tests with an internet connected testing device that reports results.

We will address each of these situations in turn. There are two aspects to result interpretation – first, does the patient understand their results, and second, are the results correctly reported to medical or public health personnel.

Self-Tests with Unreported Results In the case of routine home tests, the focus is on providing an interpretable result. There are many ways to provide result interpretation, including a visual indicator on a disposable test (e.g., pregnancy tests), reporting via an accessory device (e.g., digital glucose reading), or a result provided via a cell phone application (e.g., picture interpretation of Vessel Health app[14]). In these cases the focus is on providing a result understandable to the user and not at all reporting the results to anyone else, and any of the solutions is viable.

Self Tests Shared and Interpreted Via a Telehealth Appointment Some at home testing is planned for the near future where results are interpreted remotely (e.g., Vessel Health serology software). In this case, a diagnostic image is sent to a health care professional and analyzed during a telehealth appointment. This solves several of the issues with at-home testing by providing a way to do contact tracing (via the telehealth appointment) and helping to ensure a correct diagnostic. This enables a patient to easily understand their result and allows results to be reported to health officials appropriately. A code may be provided with each test to associate the telehealth appointment with an actual test, though there is no guarantee that the image of the test is authentic or that the person providing information to the telehealth professional is the person who actually took the test. While the latter is a problem with all self tests, the former can be addressed by using something other than just a visual image. For example, a unique RFI tag or barcode could be used with an app or an internet connected device to ensure the authenticity of the test. It should be noted that while malfeasance on the part of individuals is possible with this kind of testing, it would be challenging for malicious actors to greatly affect pandemic statistics in any meaningful way. The down side, however, is that this approach is expensive and does not scale easily due to the need for health professional involvement.

Self Tests Interpreted Via a Cell Phone Mobile Application To decrease cost compared to telehealth, a cell phone mobile app could be used to capture test results. As with the telehealth appointment, the test authenticity could be verified against an online database of test identifiers, and location information from the cell phone could be added to localize the test. As with the telehealth scenario, the app would have no way to verify that the person using the app provided the specimen, or whether the test result was authentic, e.g., not manipulated by an unscrupulous user.

[14]VesselHealth," At-home testing for COVID-19 antibodies", https://vesselhealth.com/coronavirus.html, retrieved August 2020.

Further, compromise of the app or associated cloud resources via the usual range of cyber exploits could be used for large scale corruption or manipulation of data.

Self Tests with an Internet Connected Testing Device That Reports Results The final variant is a testing device that produces the results and directly communicates via the internet to report results. In this case, the authenticity and uniqueness of the test is guaranteed at the expense of complying with necessary data and patient privacy regulations. In this scenario, as long as the specimen being tested is authentic, results are not easily forged. On the other hand, as with the phone app, compromise of the web interface or database are possible.

While security, integrity, and validation for small scale testing may be easily solved with telehealth appointments, larger scale testing will require more complex software security arrangements to provide integrity and validation of test results.

7.5 Integrity of Public Bioinformatic Databases

Numerous public bioinformatics databases have been created.[15] Researchers upload annotated data for global use and sharing with the research community. In addition to the unintentional introduction of errors into the databases, concerns have been raised regarding the intentional manipulation of the content.[16] Regardless of source, database errors can be rapidly propagated through analysis, transformation, and integration of data.[17]

While deliberately malicious modification of data contributed to public databases has not yet, to our knowledge, been detected, there may be significant motives for bad actors to do so. For instance, origins of outbreaks, which can be a political issue, can be determined from genomic sequences[18] and modification of these genomic sequences in databases may be to the advantage of those seeking to discredit other groups or deflect blame from themselves.

Preventing corruption of public databases has other important practical implications as well. For instance, sequence screening, such as that practiced by members of the International Gene Synthesis Consortium (IGSC), verifies that sequences of genes ordered by customers do not contain regulated pathogen sequences or other

[15] See http://www.oxfordjournals.org/nar/database/a/ for a partial listing.

[16] J. Caswell, J.D. Gans, et al., "Defending Our Public Biological Databases as a Global Critical Infrastructure", Front. Bioeng. Biotechnol., 05 April 2019, https://doi.org/10.3389/fbioe.2019.00058

[17] R. Pool, J. Esnayra, "Bioinformatics – Converting Data to Knowledge", National Research Council, Washington, DC: The National Academies Press, 2000. https://doi.org/10.17226/9990

[18] Liangsheng Zhang, Jian-Rong Yang, Zhenguo Zhang, Zhenguo Lin, "Genomic variations of SARS-CoV-2 suggest multiple outbreak sources of transmission', medRxiv, https://www.medrxiv.org/content/10.1101/2020.02.25.20027953v2, March 2020.

potentially dangerous sequences.[19] Thus, maintaining correct sequences for both pathogens and benign species is important.

Further, as shown in the IARPA funded FELIX program,[20] determination of whether a DNA sample shows signs of engineering is performed by comparison with a non-engineered reference sample. If public databases are intentionally polluted with engineered samples, engineering may not be detectable.

Methods and approaches used to detect and correct unintentional errors can also be used to detect intentional manipulation. Rigorous documentation of data provenance can help identify unauthorized changes. Finally, ontology-based approaches can detect inconsistencies in the data and enable data curators to address anomalies.

7.6 Defending Against Cyberattacks on Laboratory Devices

The internet of things (IoT) enables "advanced services by interconnecting (physical and virtual) things based on existing and evolving interoperable information and communication technologies."[21] The digital revolution in the life sciences has introduced smart laboratories that automate processes, link instruments and devices to a network, and offer new ways to create, store, share, and manipulate electronic pathogen and disease information. Unfortunately, malicious actors can exploit vulnerabilities arising from weak cyber and biosecurity policies and practices, as well as inadequately secured networks, networked laboratory equipment, automated systems, and electronic data and files. These vulnerabilities expose data to unauthorized access, use, disclosure, disruption, modification, or destruction and ultimately threaten data confidentiality, integrity, and availability. Cyber adversaries include both state and non-state actors.[22]

Adversaries have targeted medical and laboratory devices with malware and exploited vulnerabilities in imaging equipment, and medical and point-of-care diagnostic devices.[23,24] Additionally, it has been shown that malicious actors could use synthetic DNA sequences encoded with malware to gain control of the computers

---

[19] Gene Synthesis Consortium, "Home Page", https://genesynthesisconsortium.org/, retrieved August 2020.

[20] Adali, et al. "Integrated Decision-Making to Detect DNA Engineering in Yeast", IWBDA 2020, August 2020.

[21] International Telecommunications Union. 2012. Overview of the Internet of Things. http://handle.itu.int/11.1002/1000/11559

[22] Carlin, John P. 2016. "Detect, Disrupt, Deter: A Whole of Government Approach to National Security Cyber Threats." *Harvard National Security Journal* 7: 391–436.

[23] Department of Homeland Security (DHS). 2019. *ICS-CERT Alerts.* December 4. https://www.us-cert.gov/ics/alerts

[24] Enriquez, Jof. 2015. "Medjacking: How Hackers Use Medical Devices to Launch Cyber Attacks." *Med Device Online.* June 10. Accessed December 9, 2019. https://www.meddeviceonline.com/doc/medjacking-how-hackers-use-medical-devices-to-launch-cyber-attacks-0001

processing the sequence.[25] These vulnerabilities can change network permissions to access a device, download sensitive patient information, alter settings, issue commands, or interfere with the intended function of a device.

Any network access point provides an opportunity for adversaries to enter and compromise network components. Research demonstrates how unauthorized users have cloned radio frequency identification cards to gain physical access to laboratory facilities and used building ventilation control systems to access customer records and payment information.[26] Network security policies and practices that permit individuals to connect personal devices (e.g., phones, computers, memory cards, etc.) expose a corporate network to potential vulnerabilities, such as untrusted content, lack of configuration control, and use of location services.[27] In addition, adversaries can manipulate, copy, or destroy laboratory databases, including inventory, sequence, and disease surveillance data. In June 2019, an Iran-based internet protocol (IP) address exploit targeted exposed systems running dnaLIMS, a web-based bioinformatics system, to gain control of the computer system and further penetrate the network.[28]

In a COVID-19 environment where resources are allocated and policies made based on statistical data derived from test data, compromised laboratory data can lead to serious negative consequences in terms of under or over response. Further, exploits can be automated so that a vulnerability in one model of a laboratory device can be used to attack that device wherever it is found throughout the world.

7.7 Protection of Intellectual Property

For quite a number of years, state actors in cyberspace have attempted to steal the intellectual property of companies and government facilities of competing states. Cyber experts have identified such actions from hacker groups such as ACT10,

[25] Ney, P, K Koscher, L Organick, L Ceze, and T Kohno. 2017. "Computer Security, Privacy, and DNA Sequencing: Compromising computers wtih synthesized DNA, privacy leaks, and more." *USENIX Security Symposium.*

[26] Radichel, Teri. 2014. "Case Study: Critical Controls that Could Have Prevented Target Breach." (SANS Instiute Reading Room). Accessed January 8, 2020. https://www.sans.org/reading-room/whitepapers/casestudies/case-study-critical-controls-prevented-target-breach-35412-study-critical-controls-prevented-target-breach-35412&usg=A

[27] Simmons, Raphael. 2017. *BYOD Security Implementation for Small Organizations.* SANS Institute. Accessed January 7, 2020. https://www.sans.org/reading-room/whitepapers/mobile/byod-security-implementation-small-organizations-38230

[28] Townsend, Kevin. 2017. *Critical Vulnerabilities Found in Popular DNA Sequencing Software.* March 10. Accessed November 2019. https://www.securityweek.com/critical-vulnerabilities-found-popular-dna-sequencing-software

believed to be sponsored by the Chinese government.[29] During the COVID-19 epidemic, government institutions and commercial biotechnology companies which have been involved in the search for a vaccine or other treatments for COVID have become new targets of cyberspace theft attempts.[30,31,32]

Other attacks in the biotechnology space have come from non-state actors. Different motives have caused hacker groups to attack private biotech firms.[33,34] In one highly publicized incident, a consortium of hacker groups pledged not to attack health care providers during the coronavirus outbreak. However, a member of this consortium, CLOP, launched a ransomware attack on ExecuPharm, a U.S. firm in Vermont, claiming that while they did not attack health care providers, commercial pharmaceutical organizations were fair game.[35] CLOP went one step further, and published personally identifiable information it found on the company servers including social security numbers, some from patient studies.

A number of steps have been advocated for improving the security of biopharmaceutical companies, especially those involved in COVID-19 medical countermeasure research.[36] These include limiting patient data on servers, secure storage of backups against ransomware, and training of personnel in cyber hygiene.

Currently, both government entities as well as private corporations are involved in the development of COVID-19 treatments and vaccines involving extensive scientific data and patient information. Should these facilities be compromised, or even

---

[29] Z. Doffman, "Chinese State Hackers Suspected Of Malicious Cyber Attack On U.S. Utilities", Newsweek, https://www.forbes.com/sites/zakdoffman/2019/08/03/chinese-state-hackers-suspected-of-malicious-cyber-attack-on-u-s-utilities/#5503d1aa6758, August 2019 (retrieved August 2020).

[30] G. Lubold and D. Volz, "U.S. Says Chinese, Iranian Hackers Seek to Steal Coronavirus Research", WSJ, https://www.wsj.com/articles/chinese-iranian-hacking-may-be-hampering-search-for-coronavirus-vaccine-officials-say-11589362205, May 2020 (retrieved August 2020).

[31] C. Corera, "Coronavirus: Cyber-spies hunt Covid-19 research, US and UK warn", BBC https://www.bbc.com/news/technology-52551023, May 2020 (retrieved August 2020).

[32] D.E. Sanger and N. Perlroth, "U.S. to Accuse China of Trying to Hack Vaccine Data, as Virus Redirects Cyberattacks", NYT, https://www.nytimes.com/2020/05/10/us/politics/coronavirus-china-cyber-hacking.html, May 2020 (retrieved August 2020).

[33] D. Bukszpan, "The cyberthreat that could derail the world's race to develop a coronavirus vaccine", CNBC, https://www.cnbc.com/2020/05/12/this-cyberthreat-could-derail-race-to-develop-a-coronavirus-vaccine.html, May 2020 (retrieved August 2020).

[34] D. Winder," COVID-19 Vaccine Test Center Hit By Cyber Attack, Stolen Data Posted Online",Forbes, https://www.forbes.com/sites/daveywinder/2020/03/23/covid-19-vaccine-test-center-hit-by-cyber-attack-stolen-data-posted-online/#9c804ab18e55, March 2020 (retrieved August 2020).

[35] Z. Whittaler, "Hackers publish ExecuPharm internal data after ransomware attack",TechCrunch, https://techcrunch.com/2020/04/27/execupharm-clop-ransomware/, April 2020 (retrieved August 2020).

[36] K. Vermes, "COVID-19 Pandemic Leaves Pharmaceutical Companies Vulnerable to Cyber Criminals", BioSpace, https://www.biospace.com/article/covid-19-pandemic-leaves-pharmaceutical-companies-vulnerable-to-cyber-criminals-/, May 2020 (retrieved August 2020).

worse incapacitated, delaying the release of therapies, there would be global implications.

7.8 Discussion

In this discussion, we have illuminated a range of key cyberbiosecurity threats to infectious disease surveillance and outbreak management. While the COVID-19 pandemic has made these concerns particularly acute, ongoing strategic investments are needed to better understand, mitigate, and defend against these and similar threats. The current pandemic illustrates well the high strategic value of such public health infrastructure, and where there is value, the remote access afforded by cyber methods creates the threat that a wide variety of actors will seek advantage through cyber exploits.

Here, we have discussed what we assess to be the most near-term and high-significance concerns around core public health functions. This is by no means, however, a comprehensive view of potential issues. Experience in other domains shows that we should expect to find other potential areas of vulnerability and potential attack surfaces. Cyber threats are always evolving, and there is no reason to believe that cyberbiosecurity will be different. Likewise, similar threats are likely to obtain in other areas relevant to public health and biosecurity, such as supply chain integrity or biological effects achieved through social media manipulation. Cyberbiosecurity concerns will not go away, and are only likely to continue to increase along with increasing biological capabilities and integration with information systems. The safety of all will depend on increased attention to and investment in mitigating these issues.

8

Biosecurity and Biosafety: Policies and Practices

Andrew Jin and Igor Linkov

Novel biology technologies like gene editing and genetic engineering are creating a proliferation of breakthroughs in engineered biological systems that will change our world in areas ranging from medicine, to textiles, to energy. New developments in gene editing technologies, especially CRISPR-Cas9, have shown early signs of extraordinary potential in a variety of fields, including from basic research, applied biotechnology, and biomedical research. While the possibility of directly targeting and modifying genomic sequences in almost all eukaryotic cells could significantly improve standards of living, these technologies have the potential to pose serious biological hazards.

These potential threats have traditionally been categorized into biosafety and biosecurity concerns (Bakanidze et al. 2010). Biosafety mainly relates to the damaging effects from a biological agent that unintentionally impacts workers and the environment. On the other hand, biosecurity risks refer to the potential misuses of synthetic biology, such as bioterrorism, biowarfare or bioattacks that could derive from the genetic engineering of organisms. Because both biosafety and biosecurity share many of the same mechanisms for adverse outcomes, we must consider them concurrently. For example, a laboratory worker who was infected by a harmful virus may spread the virus in a similar mechanism as a lab worker who intentionally released a virus. Biosecurity and biosafety concerns about both the intentional and unintentional release of harmful biological agents is not a new phenomenon, but the growth of synthetic biology as a less expensive and more accessible field has dramatically changed the potential risks of synthetic biology.

The existing risk management frameworks for biological hazards were originally developed for different biosafety and biosecurity considerations, prior to the recent acceleration of developments in synthetic biology (Trump 2017). The development

A. Jin · I. Linkov (✉)
US Army Engineer Research and Development Center, Vicksburg, MI, USA
e-mail: Linkov@usace.army.mil

of biological technologies currently leaves us unprepared and vulnerable in the face of synthetic biology's potentially negative environmental consequences. Strategies of risk assessment and resilience, or the ability to prevent and to detect and recover from harmful releases of genetically engineered organisms before they permanently alter the environment, must now be developed for robust twenty-first Century biosafety and biosecurity.

The existing risk-based management strategy applied to chemical, biological, radiological, nuclear, and explosive hazards (CBRNE) is to either (a) manage threat (i.e., prevent the spread of such materials or eliminate the capacity to produce them through nonproliferation activities) or (b) reduce system vulnerability (i.e., fostering resources and protocols that harden infrastructural and social capacity to minimize the impact of such threats should they occur). Historically, these actions, derived from foundational principles of risk assessment, worked well in situations where the manipulation, use, and consequences of the hazardous materials were reasonably detectable and measurable. Further, CBRNE security principles focus more on hazardous substances that (a) are capable of development by a limited subset of actors, and (b) could be contained within a specific physical area upon release. Security policy against such hazardous substances has been one of prevention, where malicious actors are deprived of the knowledge and tools to foster harmful outcomes while others were prevented from incurring harmful accidents in their work. Given concerns of nuclear and chemical weapons as the dominating focus of CBRNE discussion, such principles were sensible for their time.

However, CBRNE strategies emphasizing the role of prevention and nonproliferation are insufficient in protecting against modern biological hazards for three main reasons. First, security threats stemming from synthetic biology are fundamentally different from those posed by chemicals, explosive, or radioactive material. For example, in the increasingly globalized and diversified field of synthetic biology, actors can use genetic engineering and editing technologies to alter or create a variety of platforms, including viruses, microorganisms, multicellular organisms, prions, and even cell-free systems (Gronvall 2019). In such circumstances, the threat in question is the engineered platform, not the passive propagation of materials through the environment. Engineered biological systems have the capability to persist, multiply, spread, and mutate beyond experimental and/or intended environmental deployment bounds, making it difficult to keep such systems under control or from interfering with the natural environment. Nuclear and chemical weapons are physically contained to the location of their use but engineered biological weapons could sweep across entire continents.

Second, exposure and vulnerability to synthetic biology threats are difficult, if not impossible, to quantify at present. We cannot confidently predict which platform might be used to generate a biological threat or weapon. Likewise, we cannot know what the target of a biological attack will be, whether it be humans, important crops, livestock, native species, the environment, or other assets. Third, it is hard to predict the consequences of release as we do not know how the weapon will be deployed against the target. The new ability to modify almost all eukaryotic cells means that any biological system could be a mechanism of disruption for such a weapon. As discussed in related literature, strategies of disruption could range from releasing an aerosolized spray containing the pathogen into the center of mass transport, to lacing the feed of cattle, to sending a computer virus encoded in genetic material to a classified laboratory in order to hack into computers upon sequencing. Even if we were able to implement systems to capture all exposures and vulnerabilities of a biological attack on a specific system, the increasingly interconnected social, technical, and economic networks with which novel biotechnologies will interact further makes risk analysis of many individual components cost and time prohibitive.

To remain in-step with the shifting biosecurity landscape, systemic capabilities for biosecurity governance must incorporate elements of resilience alongside existing tools of prevention and nonproliferation. In contrast to risk-based approaches, resilience-based approaches acknowledge that, given the broad uncertainty of the threat landscape in biotechnology, threat events are difficult to predict but inevitably will occur in some form. Figure 8.1 highlights the key differences between the traditional risk-based method and a resilience-based strategy. Unlike risk-based strategies, resilience tracks both the impact of an attack as well as the system's ability to recover.

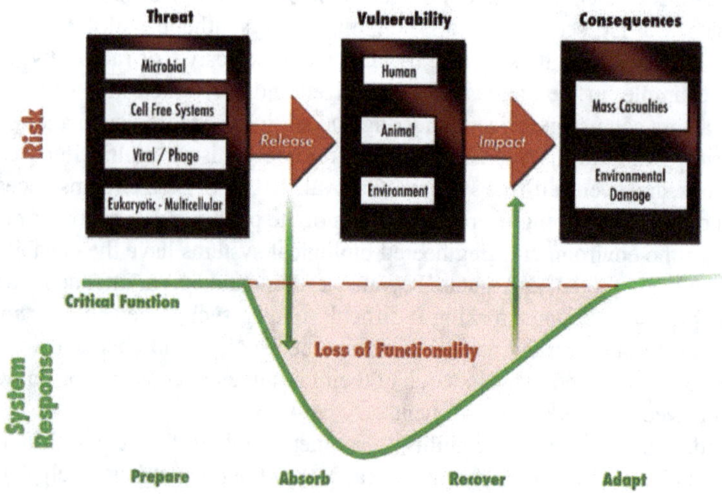

Fig. 8.1 Traditional (risk-based) and recovery-oriented (resilience-based) strategies system response. Risk-based approaches help prevent and mitigate the release of biothreats, while resilience-based strategies improve system recovery (Trump et al. 2020c)

8.1 Resilience as a Complimentary Philosophical Framework to Managing Potential Biohazards

Government officials and risk assessors are already grappling with the challenge of characterizing synthetic biology's hazards and identifying countermeasures to mitigate possible harms. This is against a backdrop of an explosion of activity in the synthetic biology landscape and increasingly inexpensive and widely available tools and techniques available to execute synthetic biology. This explosion is co-occurring with increasingly complex and interconnected global systems of infrastructure where new biological technologies will interact with systems across sectors, including the healthcare, agriculture, and energy sectors. The potential for an actor to engineer biological threats, either intentionally or unintentionally, is a real and present threat. A lack of a robust risk assessment and recovery strategy adds a further dimension of concern that may be impossible to ameliorate using existing safety and security operating paradigms (Greer and Trump 2019).

A 2012 National Academy of Sciences (NAS) report on "disaster resilience" defines resilience as the ability of a system to perform four functions with respect to adverse events: (1) planning and preparation, (2) absorption, (3) recovery, and (4) adaptation. While this definition has largely dominated resilience literature, this definition may conflate risk and resilience, which are two fundamentally different concepts. In this definition, adapt and recover are resilience concepts, while withstand and respond are concepts related to risk assessment; as such, risk is clearly added as a component of the definition of resilience. While risk assessment quantifies the likelihood and consequences of an event to identify critical vulnerabilities of a system and to harden vulnerable components of a system to avoid losses, resilience-based methods adopt a "threat agnostic" viewpoint. Thus, resilience is defined regardless of which specific threat hits the system. This is especially important for synthetic biology where new threats vary in type (e.g virus, bacterium, protozoan), source (e.g. agricultural, pharmaceutical, bioweapon, etc), and adverse implications (e.g. environmental destruction, tainted agricultural products, adverse health impact, etc.). By developing threat agnostic resilience capabilities, resilient systems can absorb and recover from biosecurity and biosafety threats quickly, regardless of the specific nature of the threat.

Resilience is a philosophy as much as a methodological practice. Resilience emphasizes the role of recovery post-disruption as much as absorption of a threat and its consequences. Resilience is grounded upon ensuring system survival, as well as a general acceptance that it is virtually impossible to prevent or mitigate all categories of risk simultaneously, and before they occur. Methodologically, resilience practitioners seek to optimize limited financial and labor resources to prepare their system against a wide variety of threats—all the while acknowledging that, at some point in the future and regardless of how well the system plans for such threats, disruption will occur. The more conventional practice of risk assessment and management is concerned with accounting for systemic threats, typically undertaken on a threat-by-threat basis in order to derive a precise quantitative understanding of how a given threat exploits a system's vulnerabilities and generates harmful consequences. Resilience complements traditional risk-based approaches by reviewing how systems perform and function in a variety of scenarios, agnostic of any specific threat.

Thus, resilience is not a substitute for principled system design or risk management. Both system design and risk management will be integral to preventing the spread of harmful biological agents into the environment, which will be key in minimizing biosecurity and biosafety risks of synthetic biology. Resilience is a complementary attribute that uses strategies of adaptation and mitigation to improve traditional risk management. Strategies to build resilience can take the form of flexible response, distributed decision making, modularity, redundancy, ensuring the independence of component interactions or a combination of adaptive strategies to minimize the loss of functionality and to increase the rate of the recovery. To address these biosecurity and biosafety challenges, risk analysis should be used where possible to help prepare for and prevent consequences of foreseeable events, but resilience must be built into systems to help them quickly recover and adapt when adverse events inevitably occur.

8.2 COVID-19 Pandemic as a Call to Action for Resilience in Biosafety and Biosecurity

On March 11, 2020, the World Health Organization officially declared the coronavirus disease 2019 (COVID-19) a global pandemic. COVID-19, an infectious disease caused by severe acute respiratory syndrome coronavirus 2 (SARS-CoV-2), is a clear indicator that we are unprepared for the potential widespread impacts of a biological hazard. The growing complexity and interdependence within various systems, such as transportation, public health, and economic systems, have made them susceptible to widespread, irreversible, and cascading failures. COVID-19 as a biological threat has shown potential catastrophic impacts of lack of resilience.

The COVID-19 pandemic is a key example where traditional risk analysis failed to accurately quantify the potential impact of a biohazard. While the possibility of a global pandemic has been a consistent concern for risk practitioners, the specific nature of the COVID-19 threat (e.g. physical symptoms, origins, rate of transmission) were likely not as predictable. Furthermore, the probability of a global pandemic occurring was not something that was cognizant on the minds of risk managers of the diverse sectors that COVID-19 would subsequently impact such as energy infrastructure, agricultural systems, or transportation. It is in such cases that developing resilient systems which can recover quickly from shocks, regardless of their sources of failure, is paramount to reducing the impact of a hazard.

Before the COVID-19 crisis, supply chains of goods and services emphasized and defined success through the efficiency in the operation, management, and outcomes of various economic and social systems, largely driven by the incentives of individual components. The COVID-19 pandemic revealed the importance of resilience within these highly complex, nested, and interconnected systems that deliver goods and services. Resilience is foundational for systems to absorb and manage shocks and prevent failures from cascading from one system component to another. (Hynes et al. 2020) For example, lack of resilience in medical supply chains to provide both preventative and diagnostic tools likely exacerbated the pandemic, causing failures to cascade into other sectors such as the food supply system(Dyal et al. 2020) and education(Viner et al. 2020). These failures elucidate the fact that mitigating hazards of a specific biotechnology, such as CRISPR-Cas9, on a specific sector requires cross-sectoral resilience to quickly respond to and absorb the adverse effects of failures before they cascade. Resilience requires an intersectoral perspective to understand key interdependencies between systems to enable them to quickly recover from adverse events and prevent cascading failures.

Given the high uncertainty and inability to fully predict or even characterize the wide universe of shocks and stresses that may challenge a given system over time, we argue that a systems theory serves as a beneficial resilience framework as it helps focus upon a given system functionality, agnostic of any given shock or stress. In other words, a systems-level approach to evaluating and enhancing resilience seeks to look internally at the structure and interrelationships of systems to review how impact or change to one node of a system generates cascading effects, in various degrees, to other directly and indirectly connected nodes. Within the next section, we explore how a resilience-based framework to adapting from the COVID-19 pandemic can benefit our understanding of biosecurity and biosafety, as these adverse events similarly entail biologically-grounded threats and uncertainties that entail implications for society at large.

8.3 A Domain-Based Resilience Framework to Adapt to Biosecurity and Biosafety

Linkov et al. (2013) assert that the Network-Centric Warfare (NCW) doctrine can be used to organize the key domains of components in a resilient system in conjunction with the four stages of the event management cycle defined by the NAS: (i) (prepare/plan, (ii) absorb, (iii) recover, (iv) adapt (Linkov et al. 2013). The NCW doctrine identifies four domains that create shared situational awareness and inform decentralized decision-making as follows:

1. Physical: sensors, facilities, equipment, system states, and capabilities
2. Information: creation, manipulation, and storage of data
3. Cognitive: understanding, mental models, preconceptions, biases, and values
4. Social: interaction, collaboration and self-synchronization between individuals and entities

All four of these are key to biosecurity in the future. By understanding resilience recourses throughout these domains, we consider the wide breadth of characteristics and decision inputs that may factor into system performance. Each domain is influenced in a different yet equally important manner when a critical or disruptive event arises, and success in one domain may not guarantee the same outcome another area. Without resilience all four components of these systems, a system cannot be resilient.

8.3.1 Physical Domain

The physical domain represents where the event and responses occur across the environment and is typically the most obviously compromised system in the midst and aftermath of an external shock or critical risk event. Physical elements can be represented by a portfolio of resilient engineering and design options that limit the exposure and damages given a biological threat. These engineering and design options include longstanding approaches of physical barriers and chemical remediation to emerging solutions related to biological engineering. They also include the development of economically feasible tools and techniques for the passive and active detection of biosecurity threats that must be developed, as early detection is an essential component in a resilience-based biosecurity policy.

While synthetic biology introduces new threats to biosafety and biosecurity, it is also an enabling technology to solutions that allow the engineering of components of biosystems that are resilient to failure. For example, Chan et al. (2016) described two microbial kill switches as "engineered safe-guard systems" to give synthetic biologists the ability to couple cell survival with a specific input signal, or to require complex environmental inputs to control circuit function (Chan et al. 2016).

The physical domain also consists of infrastructural characteristics ranging from transportation (e.g., roads, highways, railways, airports) to healthcare networks that deliver services to the public and support private business activities. As such, the physical domain of resilience thinking generally includes those infrastructural factors that are most directly impacted by a hazardous event, where the other domains include outcomes and actions that are a response to damage to physical capabilities and assets. In the physical domain, the objective of resilience analysis is to bring the infrastructural or systems asset back to full efficiency and functionality for use by its original owner or user. Recent events have shown how insufficient reserves or reliance on the global supply chain of medical supplies can make individual countries less resilient to potential biohazards when they rely on imported personal protective equipment and medical supplies (Ranney et al. 2020).

8.3.2 *Information Domain*

The information domain is where knowledge and data exist, change, and are shared. The field of synthetic biology has used open-sourcing platforms as a critical component to aid in new technology development. While these open-data resources allow for the quick dissemination of information regarding different biological systems, they also create opportunities for the dual use research of concern (DURC), in which scientific research can be used for either beneficial or malevolent purposes. In 2001, a group of Australian researchers engineered a strain of the mousepox virus, obtaining a new and more lethal strain which contributed to the understanding of poxviruses. However, this new strain could be used to engineer strains of the poxvirus that were far more lethal and transmissible for humans. (Gómez-Tatay and Hernández-Andreu 2019).

Similarly, the open-sourcing of biotechnologies opens up a wide range of concerns with "do-it-yourself" (DIY) synthetic biology. Biotechnology is becoming increasingly accessible to a larger number of people. This is especially true for synthetic biology, which aims to simplify genetic procedures and enable access to science for students. Such accessibility is exemplified by the international Genetically Engineered Machine (iGEM) competition for students and by practicing synthetic biology techniques and methods by non-professionals (DIYbio). It is unlikely that regulatory devices such as professional codes, export controls, or classification will be effective in the context of a deskilled, de-professionalized community of practitioners, nor is it likely that such regulatory devices could even cover the wide range of use cases for synthetic biology (Evans 2015).

One potential solution could be to expand information system responses in biosecurity and biosafety by learning from other DURC concerns, such as in cybersecurity. Within cybersecurity, a market has developed for the production and distribution of software exploits, with buyers sometimes paying over USD 100,000 for exploits and software vendors offering bounties for the disclosure of underlying vulnerabilities. This practice that is generating a transnational debate about control and regulation of cyber capabilities, the role of secrecy and disclosure in cybersecurity, and the ethics of exploited production and use (Kuehn and Mueller 2014).

Informational resilience during adverse events also requires quickly assembling knowledge about the threat and its recovery. However, key challenges exist in how to manage the release of data with concerns about privacy and data secrecy. For example, throughout the COVID-19 pandemic agencies have had to carefully consider questions about how to prevent reidentification of anonymized health data while balancing the need to disseminate data for researchers to improve understanding of the pandemic (Piller 2020). Big data collection and use has recently emphasized the particularization of data more so than the aggregation and generalization of data, and key questions have mounted concerning the legality, data quality, disparate data meanings, and process quality (Wigan and Clarke 2013). While the use of digitally available data and algorithms for prediction and surveillance of a potential biohazard (e.g. using cell-phone data to identifying people who have traveled to areas where the hazard has spread or tracing and isolating the contacts of effected people) will be paramount importance in the fight against the COVID-19 pandemic, it is equally important to use these data and algorithms in a responsible manner, in compliance with data-protection regulations and with due respect for privacy and confidentiality (Ienca and Vayena 2020). A key need for synthetic biology practitioners will be to manage the level of data particularization and dissemination.

8.3.3 Cognitive Domain

The cognitive domain refers to the frameworks that utilize the information and physical domains to make decisions and includes perceptions, beliefs, values, and levels of awareness, which inform decision-making. Along with the social domain, the cognitive domain is the "locus of meaning, where people make sense of the data accessed from the information domain" (Linkov et al. 2013). Such factors are easy to overlook or dismiss due to a reliance upon physical infrastructure and communication systems to organize the public in response to a disaster. However, perceptions, values, and the level of awareness of the public regarding the strategies to overcome shocks and stresses are essential to the successful implementation of resilience operations. In other words, without clear, transparent, and sensible policy recommendations that acknowledge established beliefs, values, and perceptions, even the best-laid plans of resilience will fall to disrepair. A robust accounting for the cognitive domain is particularly important for instances where policymakers and risk managers may have a disconnect with the local population, such as with international infrastructure development projects of health-based interventions.

A key need to resilience in synthetic biology is developing the appropriate operating procedures, scientific methods and tools, and analytical capabilities must be readily available to quickly identify, recover from, and remediate the release of a harmful engineered biological system (Kelle 2013). Rapid diagnostic tools are needed after a threat has been detected to absorb the threat. These diagnostic tools and approaches must be able to determine (a) the type and purpose of the genetic modification made, (b) who made the modification and where, and (c) what technology or techniques were used to accomplish the modification. The clarity provided through the answers to the above questions will allow stakeholders and decision makers to better understand the present threat and prevent similar future threats. Proposed policy strategies address some of these questions, but capabilities are still lacking.

8.3.4 Social Domain

The social domain represents interactions between and within entities involved. This includes the organization structure and communication for making cognitive decisions. Within the context of synthetic biology, this domain contains key stakeholders throughout academia, industry, and regulatory domains. It also consists of the international organizations and governments which guide synthetic biology to maturity.

Within the field of synthetic biology, many individuals and organizations are actively tackling the biosecurity challenge. Many established synthetic biology organizations, such as the iGEM synthetic biology competition, mandate that organizational leaders and judges conduct rigorous reviews of the materials and planned experiments of each team with any concerns screened for potential hazards by a commercial partner; all of this is part of the competition's guidance for participating students (Millett et al. 2019).

Another potential solution for stakeholder engagement for resilience assessment is the Responsible Research and Innovation (RRI) approach to assess societal implications of emerging research to better align processes and expected outcomes with the needs and values of society. Where top-down governance proves insufficient, other actors such as universities, non-profits, and companies will need to act as gatekeepers and watchdogs to protect against nefarious actors. Top-down governance may then support such initiatives, which will require harmonization and communication at the international level.

Within the context of adverse event response, resilience will require largescale national and international cooperation. In the context of the global COVID-19 pandemic, countries that had improved intragovernmental coordination due to past experiences with pandemic response such as 2002–03 in Hong Kong and Singapore, and the H5N1 avian influenza in 1997 in Hong Kong, were able to quickly developed plans to sustain routine health-care service, open lines of international coordination, and train staff to adhere to new prevention and control measures (Legido-Quigley et al. 2020).

The social domain also provides an area to which careful attention should be paid in overall community resilience. Social aspects of society have impacts on physical health (Ebi and Semenza 2008). Throughout both the HIV epidemic and the COVID-19 pandemic, both governmental denialism and unsubstantiated regarding home remedies that could cure or prevent illness strongly influenced adverse outcomes (Mian and Khan 2020; Linkov et al. 2021). Even with strong public health leadership, other events, such as the democratic protests in Hong Kong, can erode the public trust of authority figures. (Legido-Quigley et al. 2020).

8.4 Discussion

The domain framework outlined here ensures policymakers and risk managers acquire a holistic understanding of how a shock or stress could trigger consequences that were previously difficult to comprehend. These domains overlap highly and work together in a system. While much of the focus on synthetic biology has been on the development of biocontainment through engineering organisms, (Lee et al. 2018; Chan et al. 2016) these developments are just one portion of the physical domain that consists encompasses the larger ecosystem of biotechnologies.

Emerging technologies, like synthetic biology, often develop out of sight of social scientists and policy commentators because institutional incentives to advance science and technology usually do not create opportunities for inquiry and discussion between developers, risk assessors, ethicists, and policy analysts at the early stages of research (Trump et al. 2020b). However, a key component of developing biosecurity and biosafety will be the co-development of the social science tools and frameworks that can elucidate the ethics, morals, and risk to health, are often relegated to an afterthought and isolated within institutions or organizations. Biosecurity and biosafety resilience will require understanding the key stakeholders that make decisions in how we prepare for, respond to, recover from and adapt to stresses, and how those stakeholders utilize new information to make biosafety and biosecurity decisions.

Biosecurity policies and practices must be updated to accommodate the novel challenges associated with synthetic biology and acknowledge the globalized and diverse nature of its threat space (Trump et al. 2020a). However, biosecurity efforts remain mired in uncertainty about the capabilities of SB and its practitioners' motivations in the growing number of contexts in which it is applied. Two decades into the twenty-first century, governments are still imposing old rules on a new technology, an insufficient strategy to provide security in the future. Given that such threats can arise at any time across the globe, scientists and policymakers have a narrow and closing window in which to develop systems-level prevention and recovery-based resilience strategy for twenty-first Century biosecurity.

References

Bakanidze L, Imnadze P, Perkins D (2010) Biosafety and biosecurity as essential pillars of international health security and cross-cutting elements of biological nonproliferation. BMC Public Health BioMed Central. https://doi.org/10.1186/1471-2458-10-S1-S12

Chan CTY, Lee JW, Ewen Cameron D, Bashor CJ, Collins JJ (2016) 'Deadman' and 'passcode' microbial kill switches for bacterial containment. Nat Chem Biol 12(2):82–86. https://doi.org/10.1038/nchembio.1979

Dyal JW, Grant MP, Broadwater K, Adam Bjork MA, Waltenburg JD, Gibbins CH et al (2020) COVID-19 among Workers in Meat and Poultry Processing Facilities — 19 states, April 2020. Morb Mortal Wkly Rep 69(18):557–561. https://doi.org/10.15585/mmwr.mm6918e3

Ebi KL, Semenza JC (2008) Community-based adaptation to the health impacts of climate change. Am J Prev Med. Elsevier Inc. https://doi.org/10.1016/j.amepre.2008.08.018

Evans NG (2015) Dual-use decision making: relational and positional issues. Monash Bioeth Rev. https://doi.org/10.1007/s40592-015-0026-y

Gómez-Tatay L, Hernández-Andreu JM (2019) Biosafety and biosecurity in synthetic biology: a review. Crit Rev Environ Sci Technol 49(17):1587–1621. https://doi.org/10.1080/10643389.2019.1579628

Greer SL, Trump B (2019) Regulation and regime: the comparative politics of adaptive regulation in synthetic biology. Policy Sci 52(4):505–524

Gronvall GK (2019) Synthetic biology: biosecurity and biosafety implications. In: Defense against biological attacks. Springer International Publishing, pp 225–232. https://doi.org/10.1007/978-3-030-03053-7_11

Hynes W, Trump B, Love P, Linkov I (2020) Bouncing forward: a resilience approach to dealing with COVID-19 and future systemic shocks. Environ Syst Decis 40:174–184. https://doi.org/10.1007/s10669-020-09776-x.

Ienca M, Vayena E (2020) On the responsible use of digital data to tackle the COVID-19 pandemic. Nat Med. Nature Research. https://doi.org/10.1038/s41591-020-0832-5

Kelle A (2013) Beyond patchwork precaution in the dual-use governance of synthetic biology. Sci Eng Ethics 19(3):1121–1139. https://doi.org/10.1007/s11948-012-9365-8.

Kuehn A, Mueller M (2014) Shifts in the cybersecurity paradigm: zero-day exploits, discourse, and emerging institutions. In: ACM international conference proceeding series, 15–18-September-2014. Association for Computing Machinery, New York, pp 63–67. https://doi.org/10.1145/2683467.2683473

Lee JW, Chan CTY, Slomovic S, Collins JJ (2018) Next-generation biocontainment Systems for Engineered Organisms. Nat Chem Biol 14(6):530–537. https://doi.org/10.1038/s41589-018-0056-x

Legido-Quigley H, Asgari N, Teo YY, Leung GM, Oshitani H, Fukuda K, Cook AR, Hsu LY, Shibuya K, Heymann D (2020) Are high-performing health systems resilient against the COVID-19 epidemic? Lancet. Lancet Publishing Group. https://doi.org/10.1016/S0140-6736(20)30551-1

Linkov I, Keenan J, Trump BD (2021) COVID-19: systemic risk and resilience. Springer.

Linkov I, Eisenberg DA, Bates ME, Chang D, Convertino M, Allen JH, Flynn SE, Seager TP (2013) Measurable resilience for actionable policy. Environ Sci Technol, September:130903081548008. https://doi.org/10.1021/es403443n

Mian A, Khan S (2020) Coronavirus: the spread of misinformation. BMC Med. BioMed Central Ltd. https://doi.org/10.1186/s12916-020-01556-3

Millett P, Binz T, Evans SW, Kuiken T, Oye K, Palmer MJ, van der Vlugt C, Yambao K, Yu S (2019) Developing a comprehensive, adaptive, and international biosafety and biosecurity program for advanced biotechnology: the IGEM experience. Appl Biosaf 24(2):64–71. https://doi.org/10.1177/1535676019838075.

Piller C (2020) Data secrecy may cripple U.S. attempts to slow pandemic. Science 369(6502):356–358

Ranney ML, Griffeth V, Jha AK (2020) Critical supply shortages - the need for ventilators and personal protective equipment during the Covid-19 pandemic. N Engl J Med. Massachussetts Medical Society. https://doi.org/10.1056/NEJMp2006141

Trump BD (2017) Synthetic biology regulation and governance: lessons from TAPIC for the United States, European Union, and Singapore. Health Policy 121(11):1139–1146

Trump BD, Galaitsi SE, Appleton E, Bleijs DA, Florin M-V, Gollihar JD, Alexander Hamilton R et al (2020a) Building biosecurity for synthetic biology. Mol Syst Biol 16(7):e9723

Trump BD, Cummings CL, Kuzma J, Linkov I (2020b) Synthetic biology 2020: Frontiers in risk analysis and governance. Springer, Cham

Trump BD, Keisler JM, Volk KM, Linkov I (2020c) Biosecurity demands resilience. Environ Sci Technol 54(8):4706–4708

Viner RM, Russell SJ, Croker H, Packer J, Ward J, Stansfield C, Mytton O, Bonell C, Booy R (2020) School closure and management practices during coronavirus outbreaks including COVID-19: a rapid systematic review. Lancet Child Adolesc Health. Elsevier B.V. https://doi.org/10.1016/S2352-4642(20)30095-X

Wigan MR, Clarke R (2013) Big data's big unintended consequences. Computer 46(6):46–53. https://doi.org/10.1109/MC.2013.195

9

Types of Information Hazards and their Management

Anna Cornelia Nieuwenweg, Benjamin D. Trump, Katarzyna Klasa, Diederik A. Bleijs and Kenneth A. Oye

Biotechnology innovation has never been more accessible to individuals, companies, and (research)organizations. Advances in genetic engineering, declining costs, and improved education have improved access to biotechnologies. Such openness has provided many benefits as biotechnology has been used to address some of the world's most intractable problems However, increased access to biotechnology tools and knowledge may also pose risks to humans, animals, and the environment (Meyer 2013; Kera 2014; Li et al. 2017; Oye 2012).

Biosecurity policies seek to limit risks of misuse of biotechnology and enabling sciences (Kelle 2009). Material resources such as funding, laboratory access, possession of critical materials, and control over tools are critical components of the broader biosecurity equation but are often not sufficient. Both deliberate malevolent

A. C. Nieuwenweg (✉)
Department of CBRN Protection, Netherlands Organization for Applied Scientific Research TNO, The Hague, The Netherlands

Laboratory of Microbiology, Department of Agrotechnology and Food Sciences, Wageningen University and Research, Wageningen, The Netherlands

B. D. Trump
US Army Engineer Research and Development Center, Vicksburg, MI, USA

K. Klasa
University of Michigan School of Public Health, Arbor, MI, USA

D. A. Bleijs
Netherlands Biosecurity Office, National Institute for Public Health and the Environment, Bilthoven, The Netherlands

K. A. Oye
Massachusetts Institute of Technology, Program on Emerging Technologies, Cambridge, MA, USA

and unintentional negligent misapplications of biotechnology require access to information, inspiration, and know-how as well as material resources.

This chapter focuses on how to foster access to critical knowledge for responsible practitioners and developers while limiting information access for malevolent or irresponsible actors. An overview is given of what information hazards mean within the context of the developing field of biotechnology and synthetic biology (which is defined as "apply[ing] standardized engineering techniques to biology and thereby creat[ing] organisms or biological systems with novel or specialized functions (Issues, U.S.P.C.f.t.S.o.B 2010)), and discusses why responsible actors need to appreciate the significance of information hazards.

The first half of this chapter classifies types of information hazards, discusses under which circumstances information may pose hazards, and suggest what can reasonably be done to communicate scientific advances while limiting the potential for information misuse. Biotechnology governance requires a balance of encouragement and education on one hand, and active and passive surveillance of potentially abusable information on the other hand.

The second half of the chapter discusses practical problems associated with acting on these concerns. These tasks are complicated by shifting stakeholder and user bases, as elements of emerging biotechnologies like synthetic biology are increasingly accessible outside of conventional large governmental, university and corporate labs. Likewise, the global accessibility of biotechnology capabilities and education may reduce the ability of national governments to address information hazards from arising, especially in instances where national differences exist regarding norms, values, and ethics of what forms of biotechnology research are permissible.

Though this chapter does not provide a definitive solution to concerns over information hazards, it describes the broad problem and provides directions on how such hazards may be better monitored and addressed in the near future.

9.1 What Is an Information Hazard?

Classifying what comprises an information hazard is an inherently subjective exercise. The ability for biotechnology information to be misused is the product of not only the intellectual capacity possessed by the malevolent actor, but also their raw creativity and ability to imagine how a biotechnology or its enabling sciences might be engineered and crafted in a manner that is deliberately harmful. A manner which should at least be somewhat surprising to the broader scientific audience. On the other side of the coin, those who would surveillance and govern information hazards do so while operating under respective political, institutional, and cultural frames and incentives that bias identification and interpretation of such hazards (Lewis et al. 2019).

Though no definition or comprehensive guidance exists for biotechnology's information hazards, scientists and governments have long been concerned about how innovative or disruptive advances within various scientific fields might encourage or enable adversaries. As an operating definition, information hazards may be understood as the rate-limiting step that connects a normatively bad actor with the missing inspiration, knowledge, and processes to deploy scientific capabilities for harmful purposes (Bostrom 2011). Within such an understanding, that rate-limiting piece or pieces of information would be difficult or even impossible for the actor or organization to overcome within a set timeframe through independent research and development.

What makes the governance of information hazards so difficult is the varying forms that information hazards may take. The most intuitive example includes instances where a malevolent actor lacks some core competency or critical piece of intelligence to develop and deliver a hazardous material. Likewise, however, other cases may include malevolent actors which do have advanced basic and research capabilities to foster such threats, yet lack the inspiration, motivation, or direction to act upon that knowledge. Recent decades have largely focused upon the former example, although the latter is equally disconcerting for biotechnology given increasing levels of access to potentially dual-use scientific information to actors with limited institutional oversight from longstanding authorities.

Perhaps the most well-known exercise of governance of information hazards includes government secrecy programs, where certain scientific research was restricted in knowledge and access to approved parties only. Historically, for centuries governments have kept tight control over scientific projects that might yield a strategic advantage in military situations. They were eager to use such scientific advantages as a force multiplier against their enemies yet concerned about losing such strategic advantage should knowledge of how the technology is developed or deployed be made available to other nations. Within the twentieth century, various governments have developed and maintain information classification systems, whereby sensitive intelligence with the potential to foster harm to national security is collected and protected in a range of secure information systems.

The Manhattan Project provides a clear example of information hazards. The US nuclear weapons program in World War II rested on advances in basic and applied sciences. The very existence of the program and core discoveries and scientific breakthroughs were held closely, with no public dissemination. Personnel with access to files were carefully vetted to prevent information spillage to adversaries. Information hazards then and now include the capacity to inspire or educate potential developers. Yet important differences exist (Aldrich et al. 2008).

Today, concerns focus on how easily information might be transmitted from secure research facilities and information storage systems to foreign governments and organizations. Until recently, the primary mode of transfer for such information hazards was either through theft or copying of physical paper documents, or the acquisition of certain scientific personnel with knowledge of the specific information by means of bribery, kidnapping, or similar measures.

With the increasing maturity of high-speed Internet, prior barriers to the transfer of information hazards have been significantly degraded. The robust government classification systems continue to operate around the world, information hazard spillage is an increasing concern. Prior information hazards, such as the Anarchist's Cookbook, could spread globally through print media, yet lack the immediacy and far superior range that Internet dissemination has been demonstrated to have in virtually all countries.

Like many other emerging sciences like chemistry, nuclear physics, engineering, and computer science, biotechnology is rife with information hazards. Popular literature is riddled with examples of how a malevolent actor might manipulate a pathogen or deployed biological weapon whose consequences are sweeping and often irreversible. The difficulty of identifying and governing information hazards and biotechnology is that the core technical and scientific knowledge that might enable the creation of a weapon could also yield untold benefits to broader society (Lewis et al. 2019; Casadevall et al. 2014). This governance challenge is one of managing dual-use information, where the dissemination access to such information requires an implicit trade-off between the benefits of technological innovation to medicine, industry, and various other fields, against the potential that it might be deliberately or negligently misused. This challenge is not a new one - the example of knives might be overly simplistic yet apt, they can cut food and be useful tools in a workplace, yet also be utilized for the explicit purpose of harm to humans or animals.

Biotechnology information hazards can appear in many forms, and will undoubtedly shift as technologies continue to progress in their sophistication and accessibility. They might include the genetic sequence of a particularly high-risk pathogen, or instructions regarding how to assemble, use, and customize equipment that facilitates more precise and targeted genetic modification. Likewise, it may include the inspiration for deploying biotechnology assets and capabilities in previously unconsidered vectors or receptors of risk, the outcomes of which may have benefits and risks simultaneously.

9.2 When Do Information Hazards Matter?

If information hazards are a recurring fixture of human scientific progress, why is it that they matter so much now, and why specifically to biotechnology? As noted above, the Internet is an important consideration here, yet the advancements of biotechnology in and of itself are influencing information hazards trade-offs in a manner where the threat of a malevolent or grossly negligent actor is far more likely in the coming years than in decades past. With the continued refinement of biotechnologies like synthetic biology, information hazards have become increasingly central to biosecurity for two reasons.

First, the ability to synthesize DNA has undercut the effectiveness of physical controls on materials to keep pathogens out of the hands of malevolent and negligent actors (Oye 2012). Limits on access to pathogens on the select agents list and controls on the transportation of pathogens under the Australia Group guidelines are still of critical importance (Rappert and McLeish 2012; Kadlec et al. 1997; Danzig and Berkowsky 1997). But the ability to synthesize pathogens from information on sequences provides a pathway around physical controls. This aspect of information hazards will become more acute as gain-of-function research produces information on how to edit or modify pathogenic sequences to increase infectivity and virulence (Noyce and Evans 2018).

Second, within the past decade, a keystone achievement of genetic engineering and synthetic biology centers around the de-skilling of certain portions of biotechnology research, enabling those with far less training and experience than in previous decades to conduct advanced biotechnology exercises (Mukunda et al. 2009). Historically, a significant limitation of information hazard transference to malevolent parties included the reliance upon (a) professionals with the graduate education in a biotechnology-relevant field, and (b) considerable financial, technical, and overhead resources to execute scientific development.

To be sure, much of the advanced research that comprises synthetic biology still does require significant training, and certain exercises and tools are still inaccessible to many individuals and organizations around the world. However, this roadblock has diminished over time as the financial costs of conducting biotechnology research in certain sectors has decreased, and experimental control and efficacy has increased for those with a moderate degree of interest in trading. Likewise, as broader biotechnology becomes more globalized through improved training, sharing of information through popular media and academic publications, and global exchanges of scientific ideas and commercial products, biotechnology research is being pursued in locations around the globe, including laboratories, institutions, companies, and schools that do not have a long-standing track record of compliance and oversight with respected institutional authorities (Millett et al. 2019).

As the benefit of skilled premiums is somewhat lessened through cheaper and more accessible biotechnology research strategies, many more individuals with various backgrounds, motivations, and interests in the execution of biotechnology research are gaining entry to the field. One popular field leading the way gene editing, the scientific potential of which is influenced by the decreasing cost of genome sequencing and synthesis on one hand, as well as simplified, relatively inexpensive, and increasingly precise tools and machines to facilitate gene editing experimentation. For the most part, this development should be celebrated, for it enables the development of biotechnology in more cost-constrained settings in a manner that improves educational opportunities while also furthering scientific curiosity and development. However, this leaves government decision-makers and other key stakeholders in a precarious balancing act.

9.3 How Might Information Hazards Be Governed?

A critical challenge for the governance of information hazards includes the trade-off of: how can we best educate and train those who might assist with information hazards management without surrendering excessive knowledge that may inspire a malevolent actor in the first place?

There are few straightforward answers to inform biotechnology governments for this problem. Most likely, information hazards governance will have to be an anticipatory as well as an adaptive process, whereby potential information hazards are continuously evaluated accounting for the potential benefits if such information is further democratized against the potential threat that such information may be easily used for nefarious purposes. Examples include research into unknown human pathogens, as well as the deployment of engineered organisms to manipulate environmental conditions for a predetermined purpose (e.g., 'biomining'). In such cases, those with oversight roles and responsibilities will be required to consider, at a minimum, whether research in these fields has a discernible positive outcome that is a net improvement over conventional scientific capabilities or commercial products. Experiments requiring material from the Select Agent and Toxin List will likely always require some degree of oversight and information classification, including restrictions on how discoveries from such research are communicated.

For other experiments that do not include known agents or toxins of concern, the interpretation of what is and what is not an information hazard becomes far murkier. There are few ready-made answers, and the concern that research with information hazard potential may arise outside of government, university, incorporate labs thorough and established oversight protocols and Institutional Review Boards (IRBs) is increasing. The best advice for information hazards governance at-present therefore includes greater emphasis upon soft law mechanisms for oversight as well as increasing collaboration between top-down and bottom-up actors in the biotechnology space. In decades past, biotechnology governance has been broadly informed by operating principles and codes of conduct that, though they have little legal enforcement, have considerable influence on the norms and expectations of various actors from differing institutions. One renowned example includes the Asilomar Conference on Recombinant DNA, which was initially formed in 1975 to discuss ethical and risk-based hazards stemming from emerging biotechnology research of the day. Comprised of dozens of industry professionals from a variety of institutions, the Asilomar Conference helped to frame and anticipate future safety and ethics concerns that might arise as the field continues to develop, as well as to provide the basis for improved codes of conduct, as well as future regulatory codes and rules inspired by the precautionary principle. Further, efforts such as Asilomar help promote awareness and trust by the broader public that more effective biotechnology governance is being discussed and constructed, potentially making consumers more amenable to such products as they reach the market.

Several illustrative cases exist, were publication sparked debate among policymakers and scientists. Perhaps the most famous case, is the H5N1 gain of function experiments that focused on whether and which mutations would result in a virus capable of being transmitted to mammals (Herfst et al. 2012; Imai et al. 2012). Knowledge about the specific mutations leading to certain gain of functions can be easily misused by malevolent actors. The controversy of H5N1 gain of function experiments lead to a self-imposed mortarium by scientists (Fouchier et al. 2012) after details from two manuscripts on H5N1 were omitted on behalf of the National Science Advisory Board for Biosecurity (Casadevall and Shenk 2012). Additionally, the Dutch Government required the authors of the paper to file for an export license to prevent export of dangerous information outside of Europe (Enserink 2015).

By way of contrast, the International Genetically Engineered Machine (iGEM) competition provides a case of early and ongoing engagement with biosafety and biosecurity risks. This global synthetic biology competition engages with thousands of high school students, undergraduates, graduate students, entrepreneurs, and community laboratories annually. Student teams work closely with coaches and contest judges to ensure compliance with all legal requirements, as well as to monitor potentially hazardous or dual-use team projects. Judges and support staff are drawn from a number of institutions, including government, but with emphasis placed upon more bottom-up governance of the competition. iGEM serves as one successful example of how improved educational opportunities alongside respected and trusted authorities can both improve global education of synthetic biology and broader biotechnology research, yet simultaneously maintain oversight and awareness of interest and activities occurring outside of large government institutions, corporate offices, or academic laboratories (McNamara et al. 2014; Millett et al. 2019).

Other critical components of bottom-up information hazard governance includes responsibility and decision-making in the publication process. Globally, the publishing industry in virtually all fields has exploded in recent decades, with biotechnology as no exception. Like the democratization and globalization of biotechnology, this should be celebrated, but also reviewed for potential opportunities to improve overall governance. For emerging biotechnologies, this may include more rigorous training for editors and associate editors regarding how to identify potential information hazards, as well as to select qualified and responsible article reviewers to make an appropriate determination. This challenges wrought with the difficult quagmire, whereby journal editors and reviewers must be given greater instruction about which information hazards to look out for, as well as possible examples of them, yet simultaneously not be given enough information that would comprise an information hazard in and of itself. Academic publication remains one of the easiest ways to transfer knowledge of scientific breakthroughs globally, with higher impact in prestigious journals having tens to even many hundreds of thousands of readers on a weekly to monthly basis. Identifying and training operators of those journals that have a reasonable potential to have paper submitted to their journal that may contain information hazard is an urgent need, particularly in standardizing how potential or confirmed information hazards material should be treated within the publication process as well as without. Such measures will likely only be a low fence to prevent spillage of information hazards, yet journal editors and reviewers will likely be one of the first and possibly only lines of defense that many countries have with respect to information hazards governance on a day-to-day basis.

A long-standing challenge of biotechnology governance related to information hazards includes the construction of all offensive biotechnology capabilities (e.g., biological weapons). The 1975 Biological Weapons Convention prohibits the development, production, and stockpiling of biological and toxin weapons, including all microbial or other biological agents and toxins as well as their means of delivery. The critical distinction includes a focus on offense of use, for which all experimentation is to be prohibited. However, the Biological Weapons Convention makes exceptions for medical and defensive purposes in small quantities, leaving a window open for certain controlled experimentation to continue, provided that any quantities of biological material used and stored are justified in their permitted purpose. The BWC's restrictions have been generally successful in the decades that followed its passage, despite infamous examples of biological weapons programs continuing even after the host government signed the treaty. The important distinction for information hazards is that some research on certain biological material with a medical or defensive purpose will likely inherently have some offense of capability within the hands of a malevolent actor, making it critical to consider the level of oversight and information classification that should be applied to such cases. In addition, special consideration should be given to international research programs that include participation from one or more countries that are known BWC violators or are parties of concern, given the potential for any research or knowledge gained from medical or defensive experimentation to be utilized for other unpermitted purposes.

These and various other exercises that must come to comprise information hazards governance must take on anticipatory and adaptive governing capabilities (Esvelt 2018). Anticipatory, due to the evolutionary nature of biotechnology research, where breakthroughs enable significant departures from one school of thought and into another, thereby fostering new opportunities for information hazards and biosecurity threats to arise. Adaptive, due to the fact that risk assessors and various other decision-makers in the biotechnology governance process iteratively adjust their perception of a given biological threat over time, whereby new information is gained and best practices are adjusted to accommodate for new practices on how the deployment of genetically altered material onto humans, animals, and the environment may generate unacceptable risk. Both anticipatory and adaptive governing procedures for biotechnology within a given country must include a multi-stakeholder approach to inform best governing practices in the years ahead.

9.4 Information Hazards: Where Do We Go from Here?

To date, thankfully, there has been no major disaster pertaining to emerging biotechnologies on the scale of Chernobyl or Fukushima Daiichi. This is due in no small part to many well-trained scientists, practitioners, and policymakers that have been actively engaged with international biosecurity governance for several decades. However, as the pace of biotechnology innovation accelerates, existing hard law procedures and practices will be necessary but potentially not sufficient to fully capture shifting development capabilities and incentives in the development and commodification of biotechnology research and products.

This chapter's intent was not to solve the question of the information hazard governance dilemma, but instead to highlight why it is particularly important for emerging biotechnologies like synthetic biology. Each year, more actors become interested and involved with emerging biotechnology research, including differing cultures and governments differing perspectives on ethical, legal, moral, and risk-informed best practices for biotechnology moving forward. Navigating these intricate differences will be a considerable challenge, both for biotechnology intended for environmental deployment, as well as within human subjects' research. One conclusion that can be drawn now is that top-down hard law mechanisms alone will be generally insufficient to bridge the gap between these differing national incentives and research practices, as well as to adequately monitor the rapid growth of bottom-up biotechnology research.

Over the next several years, information hazards governance will likely become an increasing discussion point within broader biotechnology policy as well as expected practice within the lab. Continued innovation in the biotechnology space will see decreasing cost underpinning such research, such as with the dramatically decreased time and financial resources required to sequence and synthesize a genome, allowing many more actors to take advantage of such scientific breakthroughs. Future conversations about improving biotechnology governance must include improved education and awareness about such concerns in a manner that does not dis-incentivize responsible research, as well as testing and implementing anticipatory and adaptive governance bodies and operating practices that can keep pace with the accelerating rate of biotechnology innovation (Esvelt 2018; Trump et al. 2020).

References

Aldrich S, Newcomb J, Carlson R (2008) Scenarios for the future of synthetic biology. Ind Biotechnol 4(1):39–49

Bostrom N (2011) Information hazards: a typology of potential harms from knowledge. Rev Contemporary Philos 10:44–79

Casadevall A, Shenk T (2012) The H5N1 manuscript redaction controversy. Am Soc Microbiol 3(1):e00022-12

Casadevall A, Enquist L, Imperiale MJ, Keim P, Osterholm MT, Relman DA (2014) Redaction of sensitive data in the publication of dual use research of concern. Am Soc Microbiol 5(1): e00991-13

Danzig R, Berkowsky PB (1997) Why should we be concerned about biological warfare? JAMA 278(5):431–432

Enserink M (2015) Dutch appeals court dodges decision on hotly debated H5N1 papers. Science 16:89–97

Esvelt KM (2018) Inoculating science against potential pandemics and information hazards. PLoS Pathog 14(10):e1007286

Fouchier RA et al (2012) Pause on avian flu transmission research. Science 335(6067):400–401

Herfst S et al (2012) Airborne transmission of influenza A/H5N1 virus between ferrets. Science 336(6088):1534–1541

Imai M et al (2012) Experimental adaptation of an influenza H5 HA confers respiratory droplet transmission to a reassortant H5 HA/H1N1 virus in ferrets. Nature 486(7403):420–428

Issues, U.S.P.C.f.t.S.o.B (2010) New directions: the ethics of synthetic biology and emerging technologies. Presidential Commission for the Study of Bioethical Issues

Kadlec RP, Zelicoff AP, Vrtis AM (1997) Biological weapons control: prospects and implications for the future. JAMA 278(5):351–356

Kelle A (2009) Synthetic biology and biosecurity: from low levels of awareness to a comprehensive strategy. EMBO Rep 10(S1):S23–S27

Kera D (2014) Innovation regimes based on collaborative and global tinkering: synthetic biology and nanotechnology in the hackerspaces. Technol Soc 37:28–37

Lewis G et al (2019) Information hazards in biotechnology. Risk Anal 39(5):975–981

Li P-E et al (2017) Enabling the democratization of the genomics revolution with a fully integrated web-based bioinformatics platform. Nucleic Acids Res 45(1):67–80

McNamara J, Lightfoot SB-Y, Drinkwater K, Appleton E, Oye K (2014, December) Designing safety policies to meet evolving needs: iGEM as a testbed for proactive and adaptive risk management. ACS Synth Biol

Meyer M (2013) Domesticating and democratizing science: a geography of do-it-yourself biology. J Mater Cult 18(2):117–134

Millett P et al (2019) Developing a comprehensive, adaptive, and international biosafety and biosecurity program for advanced biotechnology: the IGEM experience. Appl Biosaf 24(2):64–71

Mukunda G, Oye KA, Mohr S (2009) What rough beast: synthetic biology and the future of biosecurity. Politics Life Sci 28(2):2–26

Noyce RS, Evans DH (2018) Synthetic horsepox viruses and the continuing debate about dual use research. PLoS Pathog 14(10):e1007025

Oye KA (2012) Proactive and adaptive governance of emerging risks: the case of DNA synthesis and synthetic biology. International Risk Governance Council, Geneva

Rappert B, McLeish C (2012) A web of prevention: biological weapons, life sciences and the governance of research. Routledge

Trump BD, Galaitsi SE, Appleton E, Bleijs DA, Florin MV, Gollihar JD, Hamilton RA, Kuiken T, Lentzos F, Mampuys R, Merad M (2020) Building biosecurity for synthetic biology. Mol Syst Biol 16(7):e9723

10

International Biosecurity Framework: Technical Scientific Aspects

Evan Appleton and Piers Millett

10.1 Introduction

Biosecurity is a multi-disciplinary topic that covers areas of policy, public health, economics, and science. This chapter focuses on the technical scientific aspects of the current international biosecurity framework. We discuss these technical areas in terms time horizon. We begin this chapter with review of current technology within the international biosecurity framework and discuss weakness and opportunities for further work. We then focus on near-term technical developments and imminent opportunities to strengthen the existing framework. Specifically, we break down the range of issues into biological threat prevention, detection, and response. We discuss how technical tools can assist in policy development and the engineering cycle of Design, Built, and Test. Finally, this chapter describes a 'futuring' exercise conducted by the working group that created this chapter to explore broader longer-term issues in the biosecurity space.

E. Appleton (✉)
Department of Genetics, Harvard Medical School, Boston, MA, USA

Wyss Institute for Biologically Inspired Engineering at Harvard University, Boston, MA, USA
e-mail: evan_appleton@hms.hardvard.edu

P. Millett
iGEM Foundation, Boston, MA, USA

Future of Humanity Institute, University of Oxford, Oxford, UK

10.2 Current Technology

Much recent work within our current biosecurity architecture focuses on tracking of DNA sequences that are considered to be (or may be) hazardous. This framework can be further separated into two general categories: (1) Screening for hazardous DNA sequences prior to DNA synthesis and (2) DNA sequencing and repository screening for attribution of the source of physical DNA that already exists. This framework has been developed gradually over the past ten plus years by multiple organizations. It includes policies and technical tools implemented from stakeholders across industry, academia, and government.

While this collection of approaches has been developed by multiple parties to provide coverage of the broader biosecurity area, it currently exists as a fragmented framework that has a number of clear gaps, some of which are currently being addressed by a number of parties. These gaps include but are not limited to: laborious assessment of false-positive hits from security screens, no standard guideline for implementation of a universal tool (or set of tools), and no well-defined database of established threats or potential threats.

10.2.1 DNA Sequence Screening

10.2.1.1 State-of-the-Art

Companies across the world have been offering DNA synthesis as a service for recombinant DNA biotechnology for the past several decades. This technology has improved exponentially in recent years, and currently exists at a scale where many companies are able to produce custom DNA sequences for customers on the near mega-base scale. While this capability has dramatically enabled biotechnology to make a litany of breakthroughs across biology, it has also introduced a key question to the scientific community: are there DNA sequences that DNA synthesis providers should not provide to *certain customers*? And furthermore – are there some sequences that should not be generated and provided to *any* customers?

The general answer to both of these questions by the scientific community is "yes". This answer, however, is highly conditional based on a variety of factors. While the vast majority of scientists request DNA sequences that pose no known threat, sometimes sequences of known (or potential hazard) are requested for DNA synthesis. In this situation, there is a general consensus that there are some sequences that should only be provided to qualified customers and some sequences that should not be provided to anyone. Currently, a majority of large DNA synthesis providers screen orders for the presence of viral, toxic, or otherwise known hazardous sequences to avoid sending hazardous DNA to an unqualified end user. This process typically involves lookups using local alignment tools against a reference database of hazardous sequences.

In 2009, five DNA synthesis providers founded a consortium to create guidelines for screening synthetic DNA orders, called the International Gene Synthesis Consortium (IGSC).[1] IGSC provides its members guidelines for implementing synthesis screening and gives its members benchmark tests to assure that their tools meet the current screening standards. The consortium also meets regularly to discuss any needed updates to existing protocols. It is reported that approximately 80% of global DNA synthesis is provided by currently IGSC members performing synthesis screening. However, since IGSC itself does not manage a consensus software tool or database for all members and membership to IGSC is voluntary (and not a legal mandate), there is no unified implementation of these protocols and a lot of the nuanced decision making on whether or not to synthesize a sequence is still left to individual parties. No detailed guidance exists on how to handle specific edge cases where the threat potential of a sequence can't be determined.

In parallel with IGSC activities, the US HHS published a screening framework guidance[2] for all synthesis providers to use when developing their screening tools. This guidance was made with recommendations in mind from a US National Research Council report on the feasibility of identifying DNA sequences of concern.[3] This guidance has been put to use by some non-IGSC DNA synthesis providers,[4] but as with IGSC member tools, the details of how the screening is done specifically is left to the individual parties. Other large synthetic biology groups like the International Genetically Engineered Machine (iGEM) competition and Engineering Biology Research Consortium (EBRC) have their own biosecurity considerations that sometimes employ IGSC member tools, but do not currently widely use independently built DNA screening tools or have not published their own screening guidelines.

10.2.1.2 Next-Generation Tool Development

More recently, a number of other coordinated research efforts have been initiated to fill existing gaps in the global sequence-screening framework. These have included technical research programs, screening guidance updates, and attempts to develop an open-source tool for general use across the world.

In 2018, the US Intelligence Advanced Research Projects Activities (IARPA) program announced the creation of two research programs to build new tools to identify DNA sequences that pose a safety hazard or have been engineered[5,6] The

---

[1] https://genesynthesisconsortium.org/

[2] https://www.phe.gov/Preparedness/legal/guidance/syndna/Documents/syndna-guidance.pdf

[3] National Research Council. *Sequence-based classification of select agents: a brighter line.* National Academies Press, 2010.

[4] https://bliss.jgi.doe.gov/

[5] https://www.iarpa.gov/index.php/research-programs/fun-gcat

[6] https://www.iarpa.gov/index.php/research-programs/felix

former, called Fun GCAT (Functional Genetic Categorization of Threats), aims to "to develop new approaches and tools for the screening of nucleic acid sequences, and for the functional annotation and characterization of genes of concern, with the goal of preventing the accidental or intentional creation of a biological threat." This program includes development of both new computational and experimental technologies to identify known or novel DNA threat sequences. The latter program, called FELIX (Finding Engineering-Linked Indicators), aims to "to develop new experimental and computational tools to detect engineered biological systems". Together with the technologies developed under the Fun GCAT program, this research initiative has spurned the development of new tools to identify existing and novel threats more efficiently and at a higher standard than the first wave of tools currently implemented in government, academia, and industry.

In parallel with the efforts to build better technologies for sequence screening, the IGSC issued updated guidance to DNA synthesis providers based on new information since 2010.[7] This updated guidance has provided synthesis providers with additional recommendations on how to perform their sequence screening with newer data resources and recommendations on customer screening.

Finally, there have been recent efforts to use the most up-to-date guidance and resources to develop a universal DNA screening tool that could be used by and DNA synthesizer across the world. This effort has been spearheaded by the Nuclear Threat Initiative (NTI-bio),[8] in partnership with the World Economic Forum (WEF) and the NTI Biosecurity Innovation and Risk Reduction Initiative. This group has established a working group to consider the technical requirements and implementation options for an international, common screening mechanism to prevent illicit gene synthesis. The group seeks to develop, update, and expand geographic buy-in for screening and convenes specialists from gene synthesis companies, experts from the synthetic biology community, and biosecurity professionals. The mechanism will consider DNA synthesis screening options for commercial providers, as well as producers of desktop synthesizers in the future. Central to the work of this group will be recommendations for making the mechanism common, globally accessible, and regularly updated based on advances in scientific understanding. NTI and WEF have subsequently released these recommendations for consideration during the January 2020 World Economic Forum Annual Meeting in Davos, Switzerland.[9]

---

[7] https://genesynthesisconsortium.org/wp-content/uploads/IGSCHarmonizedProtocol11-21-17.pdf

[8] https://www.nti.org/about/biosecurity/

[9] https://media.nti.org/documents/Biosecurity_Innovation_and_Risk_Reduction.pdf

10.2.2 *Attribution and Traceability*

Screening DNA sequences prior to synthesis is the first technical protective measure used to assure that potentially hazardous sequences are not manufactured for unqualified end-users. But, another set of concerns is how to track and attribute DNA materials that already have already been produced, either by DNA synthesis or DNA assembly. We refer to this problem as the problem of attribution and traceability – specifically, can we determine which organization an engineered piece of DNA came from and what its engineering history might have been. While this area is currently less developed than screening prior to synthesis, there are already some tools and guidelines that have been established for this area.

In the past couple years, at least a couple have tools have been published with identifying the lab-of-origin of engineered DNA.[10,11] There has also been a competition to test and improve these tools.[12] The general methodologies use machine-learning methods on large databases of DNA sequences to detect patterns of DNA that can be associated with certain research groups. This tends to be useful because DNA is often shared within labs and varies considerably across laboratories or even small groups within the same lab.

One difficult aspect in the attribution and traceability area is a non-perfect match between digital DNA information and physical DNA. Given that many software tools have been quickly developed in recent years with high variability in logging and tracking standards, there are perhaps many records of physical DNA in databases that might not actually physically exist anywhere. A key example of this problem is the iGEM Registry – it is deliberately open-source to allow people from all over the world to enter DNA sequence information, but many of these sequences were never actually physically built. Furthermore, in many cases, the history and real physical location of these DNA fragments might be unknown. This throws a hitch in the efforts for attribution, as it cannot be taken for granted that the data they operate on maps to real physical DNA sequences in a lab.

In parallel with tools to for attribution and tracing, updates to guidelines for synthesis providers have included considerations for customer screening. Namely, the idea is that synthesis providers should not be sending DNA materials to unqualified users or institutions. For example, in 2006, a journalist at *The Guardian* was able to obtain a fragment of the Smallpox virus (*Variola major*) via having the DNA sent to their apartment.[13] Since this initial oversight by some synthesis providers, it has been the policy of most synthesis providers to verify that DNA is only shipped to

---

[10] Nielsen, Alec AK, and Christopher A. Voigt. "Deep learning to predict the lab-of-origin of engineered DNA." *Nature communications* 9.1 (2018): 1–10.

[11] Wang, Qi, et al. "PlasmidHawk: Alignment-based Lab-of-Origin Prediction of Synthetic Plasmids." *BioRxiv* (2020).

[12] https://altlabs.tech/geac/

[13] https://www.theguardian.com/world/2006/jun/14/terrorism.topstories3#:~:text=DNA%20sequences%20from%20some%20of,short%20sequence%20of%20smallpox%20DNA.

qualified research institutions, but this is only the first step of the problem – the problem is further compounded by the fact that some researchers actively study infectious agents and develop vaccines. In this case, these scientists have a valid reason to use synthetic DNA in their studies, but this must be determined on a case-by-case basis. One additional factor that complicates issues even further in some cases is that it isn't necessarily clear what links any one institution's or researcher's are between academia, industry, or government. Between all of these factors, synthesis providers often result to making complex judgment calls about whether or not to synthesize certain DNA for a particular customer and must invest internal resources to make these decisions.

Another complex issue on the horizon in this space is the development of benchtop DNA synthesizers. These devices would allow operators to synthesize DNA in house, reducing the need to order from a provider likely to screen an order. The details of how to manage biosecurity in this case are currently being heavily debated. While the manufacturers of these machines will add software to control for illicit DNA synthesis in some way, it is unclear what methods will be completely effective. In response, some government have implemented export controls on "Nucleic acid assemblers and synthesizers, which are partly or entirely automated, and designed to generate continuous nucleic acids greater than 1.5 kilobases in length with error rates less than 5% in a single run".[14] This problem is still in the early days and the solution is far from robust.

In spite of the complexities of attribution and tracing of DNA synthesis, there is some reason for optimism in this space – those who are interested in pursuing illicit DNA synthesis have historically resorted to the methods easily detectable by simple biosecurity measures. Since biology is still done largely by hand as opposed to fully automated systems, most biologists are not even aware of their habits and will probably naïvely make easy-to-catch mistakes if they decide to pursue illicit DNA synthesis. A naïve bad actor is unlikely to reinvent the wheel, and rather, simply order near full-length pathogen sequences that will get flagged in existing frameworks. A sophisticated bad actor would be very hard to catch under current standards, but most security frameworks in other domains (i.e. cyber-security) are geared towards catching naïve bad actors. While it is important for the community to quickly develop more robust policies and tools for attribution and control of materials, there so far haven't been many examples of pressing issues of the existing framework.

[14] Australia Group. "Control List of Dual-use Biological Equipment and Related Technology and Software." (2020) https://www.dfat.gov.au/publications/minisite/theaustraliagroupnet/site/en/dual_biological.html

10.2.3 *Gaps in Existing Biosecurity Technology Framework*

While we have described the formation of a technical biosecurity framework that has partial solutions for some of the problems identified in the DNA biosecurity space, there are some key shortcomings in the existing framework. Some of the key outstanding issues include:

- High false positive rate – false alarm rates in the approximate range of ~2% of screened orders are observed and must be examined by hand
- Pathogenic genes or gene fragments embedded in non-pathogenic host sequences will have low match scores and will be missed
- Matching is slow and requires significant computational power since each order must be matched to each reference pathogen
- Black list databases of pathogens may be dangerous to share with the public since they provide bad actors with a concise like of sequences of concern
- Human genes used to interfere with the metabolism will not be captured by the current guideline (e.g. the insulin gene expressed under a strong promoter)

These problems are generally acknowledged by the biosecurity community and are often debated in ongoing meetings and conferences. Each synthesis provider currently addresses these gaps in a different capacity, and some providers are not able to address some of these gaps at all for various reasons. Furthermore, there exist some more nebulous gaps that will need to be addressed at some point, even though they are not currently pressing issues. Some of these include:

- Dissemination of viral and microbial sequences that are not currently on select agents lists, but cannot be ruled out as not dangerous in a general sense
- The use of synthetic biology in virology and gene therapy fields (generally for vaccine development)
- Economic drivers running synthetic biology might be contrary to biosecurity
- Lack of consensus and standards on which parties should be screening DNA synthesis orders
- Widespread use of bench-top synthesis screening technologies fundamentally places a majority of the responsibility of using best biosecurity practices in the customer's hands as opposed to larger institutions
- The growth of DIY synthetic biology and community labs

While there are many active conversations across the community about addressing these problems, there is no existing consensus agreement on these issues. Fundamentally, many of these issues boil down to that we do not yet have good generalizable ways to measure biological threats with clear, meticulous scientific methods. While standardization like the ISO standards[15] suggested would be nice, at the moment it is too hard to really implement.

[15] https://www.iso.org/standards.html

Generally speaking, biosecurity can't be taken for granted – it comes with costs that have to be built in, regardless of the provider. Funders or investors of synthesis technologies must support this and it is often not in today's world. It has been suggested that another route for incentivizing this technology development could be done via paying for liability protection and a regulatory framework to enforce fines and financial consequences of bad practices that result in illicit DNA synthesis. However, this idea is still far off from any form of implementation.

10.3 Imminent Opportunities for Technology Development

Thus far, we have outlined existing technical aspects of biosecurity framework and identified some gaps. We identified a set of imminent opportunities for additional technology development that would be beneficial in the short-term. We also noted that other gaps are likely to only be possible to address in the medium- to long-term future We break these opportunities into biological threat Prevention, Detection, and Response as general categories. We further disseminate these areas into more specific topics in each section (Fig. 10.1).

10.3.1 Biological Threat Prevention

Since biological engineering projects are now often done through cycles of Design, Building, and Testing, we discuss these imminent threat prevention opportunities below in that context. We identify the current approaches in these areas and suggest key ways in which new technology could be developed in these areas to strengthen biosecurity.

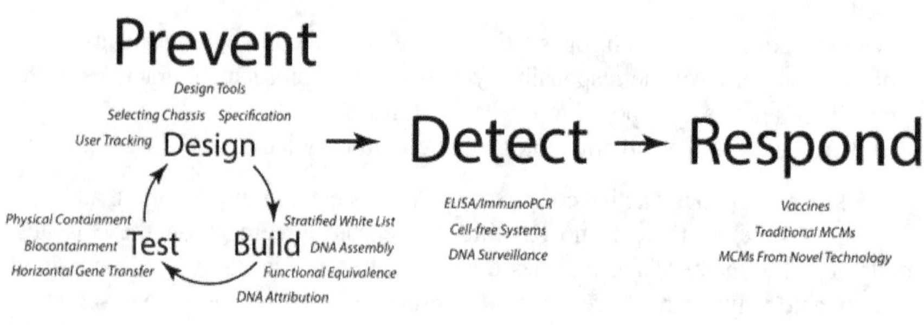

Fig. 10.1 Overview of areas for technology development in current biosecurity landscape. There is a need for technology development in all parts of the biosecurity workflow, starting from the prevention of future threat and ending with responding to bio-threats

10.3.1.1 Design

The first part of the engineering process is the Design phase. This is when the biological engineer makes key decisions about what DNA sequences will be involved in a project and which function each DNA element is supposed to have. In this starting step lies an opportunity to incorporate biosecurity features of synthetic DNA from the very beginning. In recent years, a number of software tools have been created for synthetic biology that automate the design process,[16] so there exists an opportunity to add tools specific to biosecurity in these frameworks.

Specification

The first step of a design process is an abstract design process, where the high-level design requirements are defined before a concrete design with specific real components are formulated. This step is called 'Specification'. Some examples of a specification that have been used in synthetic biology include logic gate behaviour, toggle switch behaviour, and oscillatory behaviour – for these examples, the very high level intentions of a genetic construct are defined, but no actual DNA sequences are yet selected. This step is useful because it defines the overall purpose of an engineered system before committing to the actual components needed to create a fully designed system. This specification is then generally fed automatically into a downstream design tool that chooses and arranges components to satisfy the specification.

While some software tools for specification in synthetic biology already exist, there are not yet any tools specific to specification of biosecurity features. This working group concluded that it would be useful to have such a software tool created, and discussed features that would be ideal to incorporate in such a tool. We thought that integrating security considerations into desirable system properties such as 'biocontainment' features would be desirable. We also thought it would be an opportunity to specify whether or not DNA sequences of known threat status should be incorporated into designs, or not, and create a direct link to downstream design tools.

The creation of such tools can also lead towards creation of design standards with respect to biosecurity. While our working group decided that biosecurity specification standards were a good idea, it is less clear how they would be enforced. For other specification topics in the field, complying with a specification standard is voluntary best practice. In some limited cases, like within the iGEM community, a form of project specifications with security considerations are submitted to the central organization for approval before specific designs are made, but this is not a scalable practice. In principle, a decision tree could be deduced to help an engineer

---

[16] Appleton, Evan, et al. "Design automation in synthetic biology." *Cold Spring Harbor perspectives in biology* 9.4 (2017): a023978.

determine if a specification should move forward or not, but most likely in the short term, complying with standards would have to remain a voluntary best practice.

Design Tools

After the specification step of design comes the selection of specific components to satisfy the specification. This is still most often done manually by a user in various software interfaces, but can also be done automatically with design tools. The output of this design step in a synthetic biology context is a complete DNA sequence. While current genetic design software generally does not screen the DNA sequences at this step for hazardous fragments, there is an opportunity in that step to perform an *in silico* screen in this design step for potentially hazardous sequences.

While the DNA synthesis and assembly step is a clear place to look for matches to known threats, design tools could be an ideal environment to perform modelling and analysis for potentially less obvious threat. In principle, whole-cell modelling could be developed to determine if over-expression of certain agents will disable cell metabolism via Flux Balance Analysis.[17] Models could also be developed to predict if a specific protein resembles a hazardous agent or if a viral agent could pose risk to a specific model system or cell type.

These types of modelling approaches with a complete DNA sequence could be extremely valuable at mitigating biosecurity risk before synthesis, but it is known that very few groups are capable of making such models at this time. And even then, it's hard to validate these models completely to the point that these design tools can be used reliably. Although there have been some powerful design tools recently published with highly successful design automation functionality for genetic logic circuits,[18] building tools for biosecurity threats would be a bit more abstract and require knowledge of the environment and potentially a community of cells on top of whole-cell modelling. This could be simplified for *in vitro* cell-free systems, but it is general knowledge that one cannot directly apply knowledge obtained in *in vitro* systems to *in vivo* systems.

Selecting Chassis

Once a final DNA sequence for a design is determined, it must be determined which 'chassis' (i.e. organism/model system) the DNA will be used in. Generally this information is already included in the DNA design, but when you consider the environment in which the DNA is introduced to a cell, additional considerations become relevant (especially biosecurity considerations). Furthermore, traditionally

---

[17] Orth, Jeffrey D., Ines Thiele, and Bernhard Ø. Palsson. "What is flux balance analysis?." *Nature biotechnology* 28.3 (2010): 245–248.

[18] Nielsen, Alec AK, et al. "Genetic circuit design automation." *Science* 352.6281 (2016).

some complete organisms have been designated as dangerous agents (i.e. *Yersinia pestis*) – while the organism has pathogenicity as a whole, a vast majority of the genes in these cells is harmless to humans. It would be valuable to create design tools that also consider the functionality of the chassis the DNA is implanted into – for example is it possible to add DNA to chassis that either makes a previously harmless chassis harmful or a previously harmful chassis harmless?

The simplest chassis is one that only uses cellular components, but no complete cell (i.e. 'cell-free systems'). The use of these chassis can simplify the analysis of whether an agent is harmful, but if this DNA were to get inadvertently to a living organism, it would be hard to know if it could become harmful or not. The most complicated chassis use case is a future unnatural, engineered organism or an organism that is rarely used and poorly understood in the literature. In this case, it would be near impossible to say with certainty if a DNA sequence is harmful in these contexts. This can get further convoluted if one must determine which organism or type of cell is harmed, as not all biological threats to humans are direct threats to human cells (i.e. threats to agriculture). In summary, there is ripe opportunity to develop tools that consider the organisms DNA is used in when determining DNA threat status.

Tools to Enhance Tracking of Users and Research

Finally, in addition to the many purely technical opportunities that exist for biosecurity development in the near future, there is also the opportunity to track activities at the design phase. Right now, all DNA threats are typically caught at the Build step, but if a user logs all of their design thinking, there is greater ability to warn a user of potential DNA threats before they start physical construction of the DNA. If their design was linked to databases where threat information can be automatically queried, projects that accidently use hazardous DNA components can be mitigated earlier. In the later future, if these types of tracking and logging were done at the design level, additional high-level adaptive management of new threat information could be incorporated seamlessly.

10.3.1.2 Build

After the design process, the next step of a biological engineering process is to strategize the build step. In this case, our construction material is DNA. As discussed in this chapter, there are already some existing biosecurity frameworks that relate to DNA synthesis and assembly and some gaps. In this section, we discuss some detailed near-term opportunities for technology development and technical guidelines in the Build phase.

Over the past 10 years, there has been huge technological advancement in the DNA synthesis field and has caused a shift in DNA synthesis versus DNA assembly.[19] In the past, a majority of DNA building was done in labs via PCR amplification and DNA assembly. This was necessary for building large constructs, as it was only economical to synthesize short fragments of DNA called DNA oligonucleotides (i.e. 'oligos'). However, because technology DNA synthesis of large fragments has become so much cheaper, it is now often more economical to simply outsource synthesis of most fragments and only assemble these large fragments in the last step, as opposed to relying primarily on traditional molecular cloning. As a consequence, most imminent opportunities in this space fall under the purview of DNA synthesis as opposed to DNA assembly. The working group identified a number of concrete avenues for strengthening current screening procedures and practices.

Who Should Be Screening Synthesis Production in the Future?

One general gap identified in the prior section was a lack of consensus on who ideally should be screening synthesis production. While the current system has the DNA producer (i.e. synthesis company, bench biologist, etc.) self-regulating using guidelines, as DNA synthesis becomes a more and more accessible technology, this might not remain the case. The two primary alternatives to self-regulation would be having licensed companies that provide DNA screening as a paid service or having government agencies commit resources to perform this service. Each of these alternatives has pros and cons, but both approaches generally require a centralized screening tool.

The benefit of having governments in charge of screening DNA orders is that they have a direct link to regulatory structure and no direct incentive for profit. It would also be advantageous from a centralization perspective – it could have one screening tool and one database and would not require verification that multiple screening tools and databases are screening correctly. Furthermore, a government is able to control exports of physical items (i.e. DNA). This could be a good mechanism for ensuring that hazardous DNA is not produced and exported, with legal punishment for violators. In addition, the government has access to additional information via state intelligence programs to perform customer screening (i.e. a government would have existing lists of individuals and organizations considered dangerous that it would not allow synthetic DNA to be delivered to). If done in an ideal way with neutrality towards the DNA synthesis producers and consumers, the government screening option could be a good solution. Unfortunately, in reality it might not be that simple. Currently this industry functions void of almost any regulation – adding regulation is always a messy process and each government has different attitudes towards business and science. Furthermore, a rogue state with

---

[19] Hughes, Randall A., and Andrew D. Ellington. "Synthetic DNA synthesis and assembly: putting the synthetic in synthetic biology." *Cold Spring Harbor perspectives in biology* 9.1 (2017): a023812.

control over the DNA synthesis industry could become a more general existential threat.[20]

Another alternative is having licensed companies providing screening as a service to the DNA producer. While one such company already exists,[21] more could exist in the future. Like with the government solution, this approach uses a centralized tool and database to perform screening, but is managed by a company, which has different incentives and concerns. One advantage of this solution is that it is consistent, but also is not directly tied to regulations, which allows the synthesis producers to operate with more freedom. The main complexity of this solution lies in intellectual property concerns – namely that companies have interest in not divulging their IP to other companies under the concern that the company that is screening could in principle use sequence information from the screen for economic gain. This solution requires a high degree of trust between the companies producing DNA and those screening it. This proposed solution raises all sorts of liability issues that must be negotiated between all participating parties and could be more complex than a government-based solution.

A Stratified White List Approach for DNA Synthesis Production

After a clear decision is made on the future of which parties will be performing DNA synthesis screening, a general strategy on which DNA should and should not be allowed for synthesis must be formulated. Under today's current guidelines, screeners use a Black List approach – all sequences are allowed except those that closely match DNA sequences designated as potentially hazardous. When DNA consumers make requests that are on the Black List, their orders are flagged and the DNA producer will follow up with the consumer to verify whether or not they should get the DNA. While this solution works well for sequences of known threat (i.e. fragments of the Smallpox genome), it is not able to handle new threats or threats which are not currently deemed hazardous enough to make the Black List. The result is that there is likely a large volume of sequences produced and distributed today that have some sort of threat potential. The Black List approach works well if the list remains static, but we know that in the synthetic biology space, this is not a realistic expectation.

The reverse of the Black List approach is a White List approach – a White List contains a large library (or generic definition) of sequences that cause no reason for concern. In a White List-centric approach, only things that give hits on the White List are allowed and anything not on it is not allowed. The primary difference between White- and Black-List approaches is how the 'grey' area is approached. In a Black List paradigm, things in the grey area are allowed and resources are spent to

[20] Trump BD, et al. Building biosecurity for synthetic biology. Molecular Systems Biology. 2020.

[21] https://www.battelle.org/inb/threatseq

confirm that they aren't technically on the Black List. In a White List paradigm, things that are in the grey area are generally not allowed.

This working group proposed the idea of a 'Stratified White List' approach. In this framework, it is essentially a White List approach with exceptions for highly trusted partners. These highly trusted partners could be institutions with clear approval to work with specific hazardous sequences – depending on the research being performed, there would be different categories of White Lists. The proposed category breakdown of Stratified White Lists could have the following types of breakdowns:

1. Basic molecular biology labs with institutional approval to do work in BL1 (or equivalent) with no declared intention of working on sequences that might pose threat
2. Labs with permission to work on one specific agent or set of agents with established threat status
3. Labs with permission to work on a broad set of agents with known threat status

By default, all customers (new and existing) would be automatically placed in the CATEGORY 1 provided that they have proof that they are working in an established institution (i.e. not a private address with no specific permission to work with DNA) and would then need to pass some certifications to move into CATEGORY 2 or CATEGORY 3. CATEGORY 1 would include labs at academic and industrial institutions and DIY community labs. This certification process would need to be of minimal burden to the customer, but make it clear that the customer has institutional approval to work with certain types of agents to be approved for higher categories. This process could potentially be tied to IGSC or managed in some part by a similar organization.

In general, the Stratified White List system would cover most examples of DNA to be produced, but there are a couple important edge cases that would require more thinking. First, certain mammalian genes (i.e. insulin) could be overexpressed in certain situations that make a gene on the White List cause harm to human cells. Second, this approach still does not solve the problem of sequences that are requested that match no known DNA sequence in the screening database.

While the Stratified White List approach gives a clear tiered system, one issue that could arise is how to deal with the use of middlemen or intermediary institutions giving access of higher-tiered DNA to those at a lower tier. For example, instead of a bench scientist ordering directly from a DNA producer, they might regularly order through a local supplier. Or, for example, a user at a high category clearance giving inactivated forms of agents to lower-category parties under the premise that the second party will not mutate the agent back into active form. To solve this problem, we would recommend an end-user certificate to validate that the party physically using the DNA is on the right category White List. While this proposed system does not completely solve this 'middleman problem', this problem is also unsolved in the current Black List approach.

Functional Equivalence of Sequences

The current foundation for determining if a sequence should be built or not is founded on lists of known sequences of harm. However, it is broadly acknowledged that there is a much larger list of sequences that may be threats that are not currently on these lists. This is currently unaddressed in any screening framework, but there has been discussion of methods to assign functional equivalency of sequences – the task of determining if a sequence is 'similar enough' to a known threat to cause pause before DNA synthesis.

Discussions as recently as 2008[22] had deemed this scientific pursuit too challenging of a problem to seriously consider. At that time, it was thought that a nucleotide sequence similarity percentage of 80% could be useful to identify sequences of potential threat. This sequence matching approach had many problems, namely because sequence identity isn't necessarily a good predictor of function, and was subsequently abandoned. However, in recent years, there have been huge advances made in machine learning in biology and an explosion of DNA production for genetics research. Moreover, at the time, a lot of sequence databases where new and therefore were sparse and contained errors. Now that vast, accurate databases of sequences exist and machine learning in biology has gotten off to a strong start, perhaps it is time to revisit the idea of building tools to predict functional equivalency not based solely on nucleotide sequence.

DNA Assembly and Smaller DNA Synthesis Providers

As discussed, a large shift in recent years has gone towards *de novo* DNA synthesis over traditional DNA amplification and assembly. However, DNA assembly of small fragments is still performed widely in the community too. While this DNA assembly is more time consuming and sometimes more expensive than DNA synthesis, it creates some problematic edge cases for the existing screening framework. First, since the current screening guidance only focuses on fragments of size ≥200 bp, one could order a bunch of small fragments of a hazardous agent and assemble them in a lab without being detected. Second, if a user already has access to some fragments of hazardous DNA, they can order oligonucleotides to mutate and assemble full-length agents. A near-term opportunity in this space is to build software tools that account for DNA assembly. Software could also be developed such that a DNA purchaser account could be flagged if they order a large set of small fragments that partially match a known dangerous agent or if an account suddenly logs 'unusual' ordering activity as is done sometimes with ATM withdrawals at financial institutions.

[22] National Research Council. 2010. *Sequence-Based Classification of Select Agents: A Brighter Line*. Washington, DC: The National Academies Press. https://doi.org/10.17226/12970

The DNA assembly problem is an important area to address since many types of parties still do this routinely. Organism design companies, automated platforms, cloud labs, CROs, guide RNA service providers, and other service providers regularly produce small DNA fragments in house. This issue will become even more pressing for bench-top DNA synthesizers.

Attribution and Tracing

Finally, while attribution tools have gotten off to a strong start, there are additional angles that might be factored into these tools to include lab-specific optimizations of codon optimization and synthetic biological parts usage patterns. The synthetic biology community often uses different codon optimization schemes for their parts and often re-uses combinations of characterized parts to build complex genetic circuits, so these additional dimensions could strongly aid existing attribution efforts.

In recent years, there has also been a widespread adoption of 'DNA barcoding' techniques for many areas of biotechnology and there could also be opportunities to institute a DNA barcoding system for DNA synthesis in certain capacities. This would require standardization and a consensus of how to do the barcoding, but it could be a useful way to program attribution into the DNA synthesis workflow.

10.3.1.3 Test

The final part of the Design-Build-Test cycle is Test – methods for obtaining and analysing data. While the Design and Build phases predict or assume a certain degree of functionality of an agent, in the Test phase, these qualities are scientifically determined. In the context of biosecurity, this is where the actual threat capability of any given agent is determined. In this situation, a lot of the technological focus is under what setting (both physical and biological) the testing of potential bio-threats is done, since it is dangerous to test potentially hazardous agents in an open, uncontrolled setting. Ergo, most relevant concerns with respect to testing revolve around containment.

Physical Containment

The first layer of containment is physical containment – where certain types of agents are stored and worked with by scientists. While this is conventionally performed in physically secure labs with different levels of chemical and biological agent clearance, emerging DNA technology has made this problem more complex. Specifically, with the widespread use of synthetic DNA and incorporation into model organisms, do standard decontamination and waste procedures suffice for eliminating biothreats?

In general, biological waste is either treated with bleach before being poured down the drain or sent for incineration (re-usable containers for research materials are autoclaved at high temperature and pressure). It is assumed that these procedures are broadly effective at containing biological threats, but given the stability of DNA, this generalization should be revisited in near-term research. It is known, for example, that standard sterilization techniques do not fully degrade double-stranded DNA fragments,[23] leaving whole genes unmodified. While it is not known how much DNA would be needed to create horizontal gene transfer, there is a knowledge gap of ramifications of allowing DNA to escape labs via current sterilization processes. This working group identified the area of measuring levels synthetic DNA in waste collection and the general environment as an area of key opportunity in physical containment. As an extension, another area of imminent technology development would be using new technologies to set up a surveillance network to track when DNA fragments of interest are detected at specific physical locations. Such a surveillance network could ground many of our assumptions on the physical spread of biologics from laboratories.

Biology-Based Containment

A second layer of containment is biology-based containment. Biology-based methods contain organisms using programmed biological features. The key difference in this containment approach is that it allows engineered organisms in the environment outside of a controlled facility. While this is traditionally avoided, there could be large benefit of using engineered organisms in the environment for applications such as bioremediation where organisms could be used to clean the environment of toxic molecules or pollution.

One early biology-based containment method is the use of antibiotic resistance genes such that only bacteria with that gene could grow on a substrate. Later on, other approaches of biology-based containment were developed including using cell lysates (i.e. cell-free systems) and partial organisms (i.e. lenti-viral packaging) to control biological spread by removing parts of the biology used for replication. In more recent years, technologies such as recoding,[24] kill-switches,[25] and gene drives[26] have been introduced to engineer biocontainment such that organisms can be used in certain field applications without the ability to escape the controlled environment.

---

[23] Yap et al. "Integrity of bacterial genomic DNA after autoclaving: possible implications for horizontal gene transfer and clinical waste management" Journal of Hospital Infection 83 (2013) pp 247–249.

[24] Lajoie, Marc J., et al. "Genomically recoded organisms expand biological functions." *science* 342.6156 (2013): 357–360.

[25] Stirling, Finn, et al. "Rational design of evolutionarily stable microbial kill switches." *Molecular cell* 68.4 (2017): 686–697.

[26] Esvelt, Kevin M., et al. "Emerging technology: concerning RNA-guided gene drives for the alteration of wild populations." *Elife* 3 (2014): e03401.

Kill-switch technology describes programmed mechanisms for a human observer to change the environment where an organism is placed in order to cause the organism to rapidly die. This has been engineered for both temperature and environmental triggers. Additional technology development for fine control of these mechanisms, using genetic logic gates[27] or an engineered micro-biome[28] could provide more sophisticated control of containment. Technology development in this area would have high near-term impact for biosecurity, as we think about how to introduce new, impactful biological applications, while taking proper measures to be able to control the spread of engineered organisms if they are not behaving as desired. It could also impact the desirability of biological weapons, should we develop capabilities to accurately confine engineered systems to specific locations.

A second new technology for biocontainment, called 'recoding', is a method for containing engineered agents by requiring them to use an alternate genetic code for survival. This has been done for *E. coli*, where these recoded bacteria, called *rE. coli*, require the addition of extra unnatural amino acids in the environment to survive, and thus cannot grow in environments that do not have an unnatural additive.[29] There is on-going work in the field to expand this technology into new organisms and at greater scale. Further near-term development in this area will lead to creation of organisms that are safe for use in the environment because they fundamentally cannot survive in natural environments. A key step will be experimentally demonstrating that this is true.

A third new technology, 'gene drives',[30] has been proposed as a genetic mechanism for control of population genetics. This technology uses engineered inheritance to guarantee the passing of certain genes via sexual reproduction in eukaryotes. The result is that populations could, in principle, be culled or controlled using the gene drive mechanism. The advent of this technology has drawn in large-scale science funding[31] to determine if this approach is has efficacy on a large scale and develop technologies such as reversible gene drives to correct potential mistakes made. There have also been efforts to limit the spread of gene drives to specific locations.[32,33] The primary model organism used thus far for gene drives is the mosquito, since suppression of mosquitos in certain regions could be used to supress the

---

[27] Tamsir, Alvin, Jeffrey J. Tabor, and Christopher A. Voigt. "Robust multicellular computing using genetically encoded NOR gates and chemical 'wires'." *Nature* 469.7329 (2011): 212–215.

[28] Pacheco, Alan R., Mauricio Moel, and Daniel Segrè. "Costless metabolic secretions as drivers of interspecies interactions in microbial ecosystems." *Nature communications* 10.1 (2019): 1–12.

[29] Mandell, Daniel J., et al. "Biocontainment of genetically modified organisms by synthetic protein design." *Nature* 518.7537 (2015): 55–60.

[30] Esvelt, Kevin M., et al. "Emerging technology: concerning RNA-guided gene drives for the alteration of wild populations." *Elife* 3 (2014): e03401.

[31] darpa.mil/program/safe-genes

[32] Marshall et al. "Can CRISPR-Based Gene Drive Be Confined in the Wild? A Question for Molecular and Population Biology." *ACS Chemical Biology*. 13, 2 (2018) pp 424–430.

[33] Sudweeks et al. "Locally Fixed Alleles: A method to localize gene drive to island populations." *Sci Rep* **9,** 15,821 (2019). https://doi.org/10.1038/s41598-019-51994-0

spread of malaria and other diseases. This area is ripe for additional technology development and application to more species if it can prove controllable in current research efforts.

This working group discussed ways in which the existing methods could be used synergistically to create additional layers of biocontainment. For example, one could imagine a situation where artificial dependence on certain conditions and dependence on antibiotics triggers the expression of certain factors in absence of the antibiotic to kill the cells could be a two-component system. This type of containment system could exist between an animal and bacterium where they depend on each other and one dies out in the environment without the other. Other 'xenobiotic' biocontainment examples could be developed to create complex, layered levels of biocontainment in the near future. This type of path forward will require much greater inclusion of ecologists in relevant research areas. Even then, there will be some risk in such projects. Some reversion might be possible by use of kill switches and reversal drives, but it is likely that some changes will be permanent depending not only on the system but also the population size, where it is released, and the fitness of the organism in the environment it is released into.

Horizontal Gene Transfer

Finally, in hypothetical cases of genes escaping containment, we must consider how to mitigate horizontal gene transfer. In principle, horizontal gene transfer has a certain pace with which new DNA gets introduced to a new bacterium by chance. The new DNA often has only a limited beneficial metabolic function and will for sure not be toxic to the cell. Evolution may change the DNA into genes with a more central role in metabolism or increase expression, otherwise the DNA may be lost again. In a modern global world of today and the invention of many different antibiotic drugs, the selective pressure on the bacteria has never been greater and only the most pathogenic strains can survive. Thus the acquisition of gene cassettes varies greatly.

Bacteria employ a variety of mechanisms to transfer genes horizontally, such as transformation, transduction and conjugation. Natural transformation is a process by which cells take up naked DNA from the environment. It involves multi-component cell envelope spanning structures, such as type II secretion systems (T2SS), type IV secretion systems (T4SS) and type IV pili. In transduction, DNA is transferred with the help of bacteriophages and conjugation requires physical contact between a donor and a recipient cell via a conjugation pilus, through which genetic material is transferred.

So what is transferred? A broad spectrum of mobile genetic elements, such as plasmids, transposons, bacteriophages or genomic islands are transferred and can be found to account for a large proportion of bacterial genomes as evolution goes on. An example of selective pressure is the acquisition of copper resistance (along with resistance to arsenic and cadmium) - comprising czc/cusABC and copABCD systems in the kiwifruit pathogen Pseudomonas syringae pv. Actinidae.[34] The pathogen infected the first plantation in Australia in 2010, and by 2016, 25% of all samples taken were resistant to the copper treatment.

With the development of modern molecular biology tools, endless new DNA constructs have been released into nature when biological waste is deliberately or by accident tossed down the drain. Resistance marker genes, plasmids with multi-host capabilities, and fusion proteins are a great source of DNA that can be taken up by other bacteria and which will make them even more pathogenic to human health than before.

10.3.1.4 Economic Drivers

While we have discussed here many areas of imminent technology development that could significantly bolster biosecurity practices, we must not forget the underlying economic incentives of DNA production, since economic drivers important to realising the potential of synthetic biology run counter to comprehensive biosecurity governance. Commercial applications inherently want to maximise profit and minimise overheads. Creating and implementing measures to prevent deliberate misuse add cost. This is a notable disincentive for large parts of the synthetic biology community to engage with biosecurity. Therefore, it is particularly important to streamline the financial and resource implications of biosecurity measures.

Furthermore, given its intrinsically interdisciplinary nature, many members of the synthetic biology community come from disciplines outside biology and biotechnology. As a result, they may not have been exposed to, or have a background in, biosafety or biosecurity. It is therefore important that biosecurity measures are accessible, supported by appropriate tools and resources, and adequately promoted among members of the community.

10.3.2 Detection of Biological Threats

Thus far, we have discussed numerous ways in which biological threats either are currently being mitigated or can be mitigated with technology development in the near future. However, there are also numerous opportunities to increase capability in the threat detection domain. In this domain, we assume that a biological threat has already been physically created in the environment and the question focuses on technology to detect it. In this context, we consider our ability to establish surveillance methods and rapidly diagnose biological threats.

[34] Colombi, Elena, et al. "Evolution of copper resistance in the kiwifruit pathogen P seudomonas syringae pv. actinidiae through acquisition of integrative conjugative elements and plasmids." *Environmental microbiology* 19.2 (2017): 819–832.

10.3.2.1 Diagnostics

Current methods for diagnostics of hazardous agents can be summarized as a collection of peptide sequencing, antibody-based diagnostics via ELISA or ImmunoPCR, and genome sequencing technologies. Generally speaking, peptide sequencing is most useful for protein threats like toxins, antibody screening is most often used for viral infections, and genome sequencing can be performed for both viruses and bacteria.

Of these technologies, the most rapidly evolving is genome sequencing. While some companies have made great progress on portable whole-genome sequencing,[35] there are still significant issues with the limit of DNA detection. Often there is not enough genetic material acquired in the field to make confident identifications of species and if the organism is modified, it makes that conclusion even more difficult. Some new microfluidic devices have aided this problem, but there is still a lot of room for improvement. Additional technological progress in microfluidic device development is a key area of opportunity to improve these DNA-based diagnostics. Another issue for these diagnostic devices is the comparison to reliable, non-redundant sequence databases. Historically, as large sequence databases have been built, a fair amount of inaccurate data and erroneous meta-data has been entered and so another opportunity to improve the diagnostic functionality is to clean up these databases to the point that they can be much more useful for immediate comparison with diagnostic devices.

One key area of diagnostic development for a variety of agents is cell-free systems. Cell free systems have been used for many years and are routinely produced by individual laboratories following their own recipe. Today the technology has advanced and the understanding of the important factors to make reproducible kits has enabled it to be used outside the lab.[36]

Toehold switches were developed in 2014 and utilized the preferential binding of DNA into a secondary structure if no target was present and it would unfold and bind to the target if it was present. A reporter gene would be activated upon unfolding and a signal could be detected. The technique can detect nanomolar and low micromolar concentrations due to the absent amplification step of target. It will generate signal detection in as little as 20 min and the maximum ON/OFF ratios ranged between 10- and 140-fold. Careful optimization of target region is needed to ensure maximum signal. An important advantage of paper-based distribution of synthetic gene networks is their potential for low cost (4–65¢/sensor) and relative ease to manufacture.[37]

---

[35] https://nanoporetech.com/

[36] Takahashi, M.K., Tan, X., Dy, A.J. et al. A low-cost paper-based synthetic biology platform for analyzing gut microbiota and host biomarkers. Nat Commun 9, 3347 (2018). https://doi.org/10.1038/s41467-018-05864-4

[37] Pardee, Keith, et al. "Rapid, low-cost detection of Zika virus using programmable biomolecular components." *Cell* 165.5 (2016): 1255–1266.

Developing the technique further, by using Cas9 and an isothermal RNA amplification step, the detection limit improved to low femtomolar range and had a single-base resolution discriminating between American and African Zika type viruses. Other variants of the technology (Sherlock) use an isothermal RNA amplification step to get low attomolar sensitivity by using a different Cas13 protein. This protein will, via the CRISPR methodology, find its target and cleave it.[38] Due to a build in collateral cleavage feature, the Cas13 protein will next degrade any mRNA it may find. Thus the provided fluorochrome - quencher reporter mRNA molecule will be degraded and a signal can be measured using a fluorescent reader. Yet again, other systems using cas12 can target DNA in the same way as mentioned above[39] or by using a CRISPR-Cas9-triggered nicking endonuclease-mediated Strand Displacement Amplification method named CRISDA.[40]

A technology to be fully developed in the future is the biological transistor. It is the detection of an unamplified target gene via CRISPR–Cas9 immobilized on a graphene field-effect transistor. An electrical signal is generated if CRISPR detects its target and thus can positively identify a biological agent on the DNA level within 15 min.[41]

With the invention of the methods mentioned here, a field deployable paper stick technology will be able to tell if a dangerous pathogen is present in a fast and reliable way. It is a huge step in the direction of being able to detect a biological attack on site, but still laborious work is still needed to extract nucleic acids from each sample and ID RNA is present, RNases are to be avoided at all cost to get reliable results.

10.3.2.2 Surveillance

Finally, as diagnostic tests are developed to be faster and more accurate, we can start to form systematic surveillance protocols. This can range from detection of immediate human health pathogens to analysis of field micro-biomes, detection of fungi and decomposers, and agricultural pathogens. While these other types of threats are currently too low priority to focus diagnostic efforts on, in the big picture, these areas really matter. In agriculture, there are already systems for tracking and regulating pedigree of lines of animals and plants, but as the technology for diagnostic DNA tests improves, it would be reasonable to develop the areas of

---

[38] Gootenberg, Jonathan S., et al. "Multiplexed and portable nucleic acid detection platform with Cas13, Cas12a, and Csm6." *Science* 360.6387 (2018): 439–444.

[39] Chen, Janice S., et al. "CRISPR-Cas12a target binding unleashes indiscriminate single-stranded DNase activity." *Science* 360.6387 (2018): 436–439.

[40] Zhou, Wenhua, et al. "A CRISPR–Cas9-triggered strand displacement amplification method for ultrasensitive DNA detection." *Nature communications* 9.1 (2018): 1–11.

[41] Hajian, Reza, et al. "Detection of unamplified target genes via CRISPR–Cas9 immobilized on a graphene field-effect transistor." *Nature biomedical engineering* 3.6 (2019): 427–437.

genetic surveillance of agriculture since it is a high-impact area of human wellbeing that is not directly human health focused.

10.3.3 Threat Response and Countermeasures

Biological countermeasures are typically biologics and small molecules used to detect, prevent, or treat biological and chemical insults. Biologics are composed of vaccines and antibodies. Vaccine development, while slow and laborious, is effective at producing acquired immunity and protection to a broad range of known diseases and weaponized agents. Recently, large-scale mining of human immune repertoires for antigen binders has been propelled by technological advances such as next generation sequencing (NGS) and given rise to the field of system-immunology. Coupled with bioinformatics analysis, we have gained significant insight into the diversity of antigen binders and the polarization of repertoires in response to challenge. Subsequently, it is now possible to mine these repertoires for protective monoclonal antibodies and deliver effective countermeasures. However, extant antibody discovery platforms suffer from a multitude of disadvantages that impede high-throughput repertoire interrogation and antibody discovery.

There are several large existing programs to develop medical countermeasures (MCMs) to new and existing biothreats such as p3, PRISM, and US AID, but these programs are beyond the scope of near-term biosecurity efforts to improve. While new synthetic biology tools will certainly lead to decreased development time to countermeasure delivery compared to traditional approaches, it is still relatively new technology and will take some time to be integrated into these large existing efforts to develop countermeasures.

One topic discussed at some length by this working group was how information on MCMs should be disseminated to the general public. Specifically, we discussed the idea of putting the latest technical information on new threats online. For example, each year there is a seasonal strain of influenza that circulates and vaccines are routinely developed to combat the new strain. To do this, the new viral strain is sequenced and a new MCM is created. Should these new viral sequences and information on countermeasures be publicly available information?

Historically this type of information has been available only to those actively working in the space. New sequence information is kept on a non-public database and companies that develop yearly MCMs for influenza get the physical strain in exchange for donating some vaccine free of charge for use in resource limited settings. However as they can access the data with the virus sequence, they do not have to ask for the strains anymore – the can simply synthesize the new strains themselves, make the vaccines from that source without the obligation to make the MCMs more generally accessible. Thus, the system relies on the good will of the companies to share benefits from their work.

Furthermore, we have seen a much more open approach in response to the global COVID-19 pandemic – the DNA sequence for this threat[42] and subsequent diagnostic and countermeasure development has been rapidly published and made publicly available.[43] This has led to rapid development of novel diagnostics and MCMs by a large swathe of companies (including some synthetic biology-based companies), giving tremendous opportunity for the biotechnology industry. But, the cons of this very open approach will take time to play out, as some risk has been taken by disseminating so much of this information in a short period of time to mitigate a global crisis. Only after the dust settles with the on-going COVID 19-pandemic, will we be able to see whether or not there are clear negative consequences of this open approach.

10.4 Long-Term Opportunities for Technology Development

The discussion in this chapter has thus far discussed the existing technical biosecurity framework and various areas for near-term improvement. Here, we have some additional discussion of long-term opportunities for improvement that are still too nebulous to attempt to mitigate with cutting edge technologies in the short term future.

10.4.1 Implications for Adaptive Risk Management Framework

Several aspects of technologies and products derived from synthetic biology create a need for an adaptive approach for risk assessment.[44] Synthetic biology approaches are foundational, being applied to many different problems/areas and accessible to a wide range of people. Because of this, there is great uncertainty about what type of technologies might develop from synthetic biology. At the same time, new products and capabilities are being developed at a rapid pace. As a result, it is difficult to anticipate the potential threats posed by synthetic biology using a prescribed or static approach for assessing risks and bio-threats.

---

[42] Andersen, Kristian G., et al. "The proximal origin of SARS-CoV-2." *Nature medicine* 26.4 (2020): 450–452.

[43] Thao, Tran Thi Nhu, et al. "Rapid reconstruction of SARS-CoV-2 using a synthetic genomics platform." *BioRxiv* (2020).

[44] Oye, K. "Proactive and adaptive governance of emerging risks.

The case of DNA synthesis and synthetic biology." Public Sector Governance of Emerging Risks, International Risk Governance Council (2012) https://irgc.org/wp-content/uploads/2018/09/FINAL_Synthetic-Biology-case_K-Oye_2013.pdf

We were able to identify desirable features of an adaptive risk management framework. First, an adaptive risk management would need to use widespread monitoring and data collection practices. This would likely include large-scale measurement of known hazardous DNA and engineered DNA. Secondly, there would need to be a systematic process for horizon scanning to identify new things that realistically might need to be measured. It is not feasible to have a single static list of concerns in a world where biotechnology is constantly evolving. Such a system could look a little like monitoring data from subway filters[45] with indicators of background levels of specific agents. First such backgrounds would need to be clearly established and there would have to be a mechanism to flag the presence of new, unknown particles.

10.4.2 Imagining the Far Future

Finally, the group carried out a future-casting exercise where participants imagined what the world might look in the long-term future (25–50 years). They considered the role synthetic biology (and its products) could play in that world. Such a future could be positive or it could be negative. This exercise involved scenario development, group work to identify specific biosecurity implications, and then back-casting to identify key technological developments that might be important for shaping the future.

Possible future applications of synthetic biology identified included:

- Enhanced diagnostics, such as those based on a comprehensive understanding of life processes
- Cheaper, faster and improved medicines (including at point-of-care), such as personalised medicine, smart wound healing materials, antibiotics, vaccines, other biologics (including to balance the metabolism)
- Human enhancement, such as brain/visor interfaces, augmented limbs, synthetic organelles to correct natural deficiencies (such as delivering insulin for diabetics) or communication with animals
- Microbiomes engineered to be self-balancing, offer improvements to performance, health or a more robust immune system, or more efficient in processing and absorbing nutrients from food
- Adjusted plants, resistant to challenges from climate change and food shortages resulting from expanding human populations
- On going, near-real-time remote monitoring of health and the environment
- Chemically-induced exercise through the diet
- Enhanced food, including with a long shelf life

---

[45] Grass, David S., et al. "Airborne particulate metals in the New York City subway: a pilot study to assess the potential for health impacts." *Environmental research* 110.1 (2010): 1–11.

- Improvements to the transport infrastructure, such as self-repairing or better road surfaces
- Changes to energy use, such as by increasing use of bio-lights
- Greater control over the environment, such as increasing biological diversity, through bioremediation, systems for reclaiming or making use of waste, controlling or removing pests and disease vectors, or perhaps weather control
- Cyber-genetics and enhanced immune surveillance
- Enhanced materials, such as materials that adapt to the environment, alternatives to, or new ways to recover rare earth metals
- Novel production capacities, such as directly from carbon and water, possibly via distributed production, to make things like food or high-value chemicals
- Point-of-use (distributed) manufacturing, such as through the use of well-characterized and reliable chassis organisms.

The group then split into three sub-groups based on common components of their visions of the future. The sub-groups then discussed relevant biosecurity issues, focusing on ways that a lack of, or adequate, biosecurity could have a notable impact delivering a desirable future. Intervention points identified included:

- The impact of business models in shaping integration of biosecurity and the importance of engaging the private sector on biosecurity early in technology development
- The importance of improved risk assessment and management of biosecurity risks to enable ecological or environmental uses
- Adaptive, flexible, proactive, and cooperative approaches to governance
- Involvement of publics and industry in biosecurity governance and decision-making
- Perception issues around military funding of synthetic biology
- Potential for subversion of neuro-electric interfaces, the importance of encryption, and the potential for cyber attacks to have a biological impact and to hack biological components to have a digital effect
- The possibility of biosecurity concerns to unjustifiably restricting progress in key fields (opportunity costs)
- The value of developing new capabilities to carry out applied risk assessment between the bench and field trials (pilot scale), such as through the use of isolated environments that can provide a baseline for studies

In the same small groups, participants then considered which technical interventions could help address the biosecurity issues identified. These included:

- Rapid dynamic adaptive approach, surveillance system for environment
- Technology to identify the origin of a molecule or tag DNA bases (e.g. fertilizers have to be tagged - > biological tags like isotopes)
- 'Biological encryption' to counter bio-hacking of neurons

In summary, the working group came up with a wide variety of potential future technologies and their related biosecurity risks. While most of the future technologies indeed had at least one clear biosecurity risk, the group had a mostly positive outlook on the future of synthetic biology and technical biosecurity.

11

Genetically-Altered Micoorganisms and the Soil System

Robyn A. Barbato

Microorganisms serve as agents for synthetic biology. It certain instances, the technology is reliant on survival of the genetically-altered microorganism in the natural environment. This chapter offers insight into the soil system to improve the performance of genetically-altered microorganisms in the natural environment. This chapter covers the soil system, the fitness of genetically-altered organisms, and considerations for their survival and proliferation in nature. This chapter does not discuss the ecological consequences of survival in the natural environment or the regulatory considerations to distribute a genetically-altered organism in the environment. In addition to regulatory approval, both considerations should be critically reviewed before a genetically-altered organism is distributed in the environment.

11.1 The Soil System

11.1.1 Soil Properties

The soil environment is complex and heterogeneous. Soil is rock that has been broken down and weathered and is combined with organic matter which supports growth of organisms. The *soil profile* is composed of layers, termed horizons, which are distinct with depth from the soil surface. The O horizon generally comprises leaf litter. Because of the presence of plant roots and soil microorganisms, the top layer of the soil is the A horizon, which typically harbors a high amount of organic matter. Beneath that horizon is the B horizon, which can harbor materials that have leached

R. A. Barbato (✉)
US Army Corps of Engineers Engineer Research Development Center Cold Regions Research and Engineering Laboratory, Hanover, NH, USA
e-mail: Robyn.A.Barbato@erdc.dren.mil

from the A horizon above it. The C horizon is located below the B horizon. The C horizon contains rock fragments that have partially weathered.

Soils are also laterally heterogeneous across the landscape. Soil forming factors influence the soil type present at a given location. In an attempt to characterize soils across the landscape, which includes their depth profile, the United States Department of Agriculture developed twelve orders of soil taxonomy, which include alfisols, andisols, aridisols, entisols, gelisols, histosols, inceptisols, mollisols, oxisols, spodosols, ultisols, and vertisols.

The texture of the soil is determined by the proportion of three different types of soil particles, sand, silt, and clay, which vary by their diameter. Sand particles have a diameter ranging from 0.05 mm to 2.0 mm. Silt particles have a diameter of 0.002 mm to 0.05 mm. Clay particles have a diameter of less than 0.002 mm. The surface area of these mineral particles is lowest for sand and orders of magnitude higher for clay mineral particles. Clay minerals in particular are influence biological and chemical reactions occurring in the soil, mainly due to their small size and surface charge. The negative charge of clay minerals is an important factor in the soil cation exchange capacity, or the ability of the soil to exchange cations.

The combination of sand, silt, and clay particles determine the soil texture. There are 12 major soil textural classes that are identified in the soil textural triangle. While the texture of the soil can be informative for the combination of mineral particle types, the orientation of these particles in three-dimensional space is more meaningful for the resident microorganisms. The microorganisms live in the soil pores between the soil aggregates. Soil texture is an important factor in the location and type of soil pores present. Soil pores are spaces between the soil aggregates that may be filled with water, air, organic matter, microorganisms, and nutrients (Fig. 11.1). Commonly pores include macropores (greater than 75 μm), micropores (5–30 μm), and nanopores (<0.1 μm).

Important soil abiotic attributes include soil pH, soil temperature, soil water content, and soil cation exchange capacity. Soil pH is very important for how chemicals behave in soil and which organisms grow in or on soil. Often, forest soils are acidic,

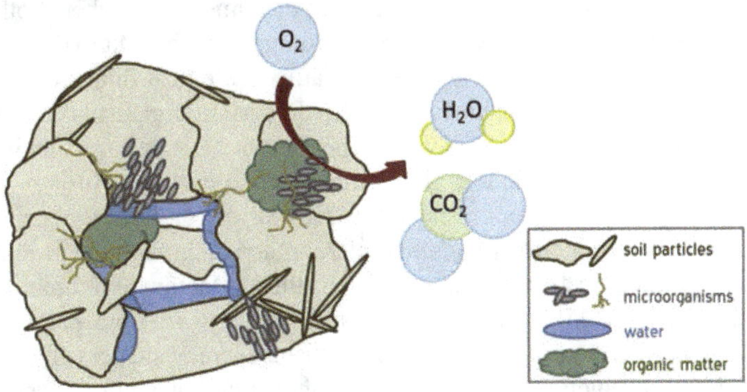

Fig. 11.1 Conceptual model of a soil pore emphasizing soil microorganisms' role in heterotrophic respiration

while soils located in arid regions are alkaline. Soil temperature influences the activity and rates of processes of organisms in soil. It is highly dependent on the amount of soil water due to the specific heat of water. Soil water potential energy of water held in the soil in relation to the energy of free water. Soil water potential includes matric potential (influence of capillary forces on the system), osmotic potential (influence of solutes on the system), and gravitational potential (influence of gravity to pull the water toward the earth's core). Soil water potential is affected by soil texture. A water release curve is often developed for a type of soil to relate the gravimetric or volumetric water to the matric potential. Soil exchange capacity is the ability of the soil particles themselves to hold either cations or ions. The clay particles influence the soil cation exchange capacity, which in turn influence the types of chemicals which are available to nearby microorganisms.

11.1.2 Life in Soil

Soil is teeming with life and supports the growth of plants and animals. Within soil, there are microorganisms at a range of trophic levels which include as viruses, bacteria, archaea, fungi, and protists. At a given moment, approximately 10% of soil microorganisms are active. The remainder are in a resting metabolic state. There are approximately one billion bacteria in one gram of soil, which includes 1000 different species. Microorganisms reside in the soil pore space. Important soil properties influence the number and types of microorganisms found in the soil.

Microbes live together in soil and do not exist alone. Therefore, ecological interactions between soil microorganisms determine their survival and proliferation. Important ecological interactions in soil include commensalism, mutualism, symbiosis, competition, amensalism, parasitism, predation, and neutralism. The positive ecological interactions are when at least one group benefits from an interaction. Positive ecological interactions include commensalism, mutualism, and symbiosis. In commensalism, one microorganism functions and another microorganism gains from that function. Mutualism is a positive ecological interaction where both microorganisms benefit. Symbiosis is where two microorganisms benefit to the extent that they cannot survive without one another. Negative ecological interactions include competition, amensalism, parasitism, and predation. Microorganisms compete over substrates for growth and/or metabolism. They might be superior in their genetic capacity to use a substrate and/or physically faster to access it through motility. Amensalism is a negative ecological interaction where one microorganism suppresses the growth or function of another microorganism through the production of compounds. A classic example of amensalism is the production of antibiotics by the fungus *Penicillium* to kill nearby bacteria susceptible to it. Both parasitism and predation include feeding on living microorganisms thereby suppressing the host's or prey's ability to function. Neutralism is an ecological interaction where neither microorganism benefits from the interaction. These interactions are unlikely in soil and difficult to define in a diverse, heterogeneous environments.

Life in soil is driven by the active microorganisms present, their access to substrates, and their genetic capacity to make enzymes to break down the substrates. Soils are generally oligotrophic, which means that they are low-nutrient systems. Depending on their geographic location, they may experience vastly different conditions throughout the day or the year. For instance, precipitation could dramatically change the water potential of a soil and in turn either create anoxic conditions and/or cause a release of substrates from one soil pore to another. Diurnal temperature cycles could span beyond the temperature optimum for resident microorganisms. Therefore, groups of microorganisms develop survival strategies to thrive under a particular set of conditions. The development of new tools to survey microorganisms will enable understanding of the spatial and temporal effects on microbial community composition and function in soils. The Earth Microbiome Project encouraged the collection of microorganisms in the field and offered a repository for the number and types of microorganisms present at a given location at the time of collection (Gilbert et al. 2010, 2014). Concurrently, soil samples are collected to culture microorganisms in the laboratory. The vast number, types, metabolic preference, and growth requirements of the microorganisms make culturing challenging. Nonetheless, it is an important step to understanding the range of metabolism and growth of the organisms.

11.2 Fitness of Genetically-Altered Organisms

11.2.1 Fitness of Domesticated Microorganisms

There have been recent advancements in the field of synthetic biology that resulted in new genetically-altered organisms. Often, microorganisms such as *Escherichia coli* (a bacterium) and *Saccharomyces cerevisiae* (a yeast) are popular targets for genetic manipulation. They are common chassis, which physically house genetic components and provide resources (i.e. transcription and translation machinery) for the organism to function (Danchin 2012; Brophy et al. 2018; Westmann et al. 2018; Trump et al. 2018). This is largely driven by the depth of knowledge and research on these organisms. They have been domesticated in the laboratory through multiple enrichments and manipulations. There is a wealth of information how their genetic material encodes for specific metabolism. Control of the growth and metabolism of these microorganisms through synthetic biology tools is extraordinary. These organisms have met the criteria identified to be an effective chassis which include that the organism grows easily, has a robust cellular envelope, transcription and translation can be controlled, their growth and metabolism is well understood, and they have low mutation rates (Foley and Shuler 2010).

11.3 Considerations for Survival in Soil

11.3.1 Soil Conditions Influence Growth

Life in soil is stressful. The soil properties described earlier in this chapter influence the composition and function of soil microbes. Soil microorganisms are reliant on either narrow or broad conditions to proliferate and metabolize substrates. Soil properties such as water potential, organic matter content, particle size distribution, and temperature influence biological activity and collectively define the state of soils. Soil temperature influences soil activity, with warmer soils typically experiencing higher rates of processes (Lloyd and Taylor 1994; Hanson et al. 2000; Fierer et al. 2006). Soil moisture also influences soil activity (Fierer et al. 2003; Schimel et al. 2007; Xiang et al. 2008; Borken and Matzner 2009; Fóti et al. 2014). The optimal matric potential for microbial activity in is −33 kPa. When the soil is too dry, soil activity is reduced or stops completely (Manzoni et al. 2012). At this point, substrate diffusion is limited, and therefore there are no substrates bioavailable for nearby soil microbes. When a soil is too wet, the diffusion of oxygen to aerobic microorganisms is limited. Concurrently, anaerobic microorganisms (those which use compounds other than oxygen as electron acceptors) become active.

11.3.2 Microbial Adaptations

Microorganisms have adapted and continue to adapt to the soil conditions. Therefore, ecological niches defined by ecological interactions and physiochemical factors are occupied. Disturbance events such as physical soil manipulation or the addition of a chemical often change the niches available to groups of microorganisms. Absence of a disturbance, the ability of an introduced microorganism to gain a foothold in the soil system is less likely. Often, the introduced microorganisms are not adapted to the conditions and are quickly outcompeted by nearby microorganisms. Unless they fulfill a key ecological role, they are likely diluted out of the soil system until they become extinct.

11.3.3 Fitness of Microorganisms in Soil

As synthetic biology advances, important considerations need to be considered to ensure optimal performance in the natural environment. Domesticated organisms likely lack fitness under the stressful conditions of soil, as soil conditions often do not meet the metabolic needs of these organisms and indigenous soil microorganisms are more adept to acquire substrates to survive. Though genetic expression and regulation of synthetic circuits are highly host-specific, the best candidates for fieldable chassis are likely environmental microbes that have adapted to survive under dynamic conditions (Tanner et al. 2017). In doing so, genetic engineering augments the process that natural selection has already optimized.

The high diversity in soil offers the opportunity to develop new chassis capable of surviving in the natural environment. There have been recent advances is the phylogenetic classification of soil microorganisms. Using sequencing technologies, the taxonomy of microorganisms in a given soil can be identified. Through culturing and phenotypic characterization, discoveries of novel microorganisms are possible. Because these microorganisms are originating from the environment they will be returned, they may be more adept to surviving the challenging conditions common to most soil environments. This implies that they retain the genetic advantages for survival under dynamic physical and chemical conditions.

11.4 Conclusions

The soil system is complex and heterogeneous in space and in time. Soils are highly diverse and present a source and a sink for new products for synthetic biology. Important considerations for survival in soil have been outlined in this chapter. Furthermore, these concepts can be extended to situations where genetically-altered microorganisms are added to the soil matrix to perform a specific function. Because soils are so heterogeneous, a singular approach is unlikely to be successfully applied to multiple soil systems. Understanding soils and the multitude of ecological interactions occurring within them is key to promote the survival of introduced organisms to that particular environment.

References

Borken W, Matzner E (2009) Reappraisal of drying and wetting effects on C and N mineralization and fluxes in soils. Glob Chang Biol 15(4):808–824

Brophy JAN, Triassi AJ, Adams BL, Renberg RL, Stratis-Cullum DN, Grossman AD, Voight CA (2018) Engineered integrative and conjugative elements for efficient and inducible DNA transfer to undomesticated bacteria. Nat Microbiol 3:1043–1053

Danchin A (2012) Scaling up synthetic biology: do not forget the chassis. FEBS Lett 23586:2129–2137

Fierer N, Schimel JP, Holden PA (2003) Influence of drying-rewetting frequency on soil bacterial community structure. Microb Ecol 45(1):63–71

Fierer N, Colman BP, Schimel JP, Jackson RB (2006) Predicting the temperature dependence of microbial respiration in soil: a continental-scale analysis. Glob Biogeochem Cycles 20(3):GB3026

Foley PL, Shuler ML (2010) Considerations for the design and construction of a synthetic platform cell for biotechnological applications. Biotechnol Bioeng 105(1):26–36

Fóti S, Balogh J, Nagy Z, Herbst M, Pintér K, Péli E, Koncz P, Bartha S (2014) Soil moisture induced changes on fine-scale spatial pattern of soil respiration in a semi-arid sandy grassland. Geoderma 213:245–254

Gilbert JA, Meyer F, Jansson J, Gordon J, Pace N, Tiedje J, Ley R, Fierer N, Field D, Kyrpides N, Glöckner FO (2010) The earth microbiome project: meeting report of the "1 st EMP meeting on sample selection and acquisition" at Argonne National Laboratory October 6 th 2010. Stand Genomic Sci 3:249–253

Gilbert JA, Jansson JK, Knight R (2014) The Earth Microbiome project: successes and aspirations. BMC Biol 12:1–4

Hanson PJ, Edwards NT, Garten CT, Andrews JA (2000) Separating root and soil microbial contributions to soil respiration: a review of methods and observations. Biogeochemistry 25:1537–1546

Lloyd J, Taylor JA (1994) On the temperature dependence of soil respiration. Funct Ecol 8(3):315

Manzoni S, Schimel JP, Porporato A (2012) Responses of soil microbial communities to water stress: results from a meta-analysis. Ecology 93(4):930–938

Schimel J, Balser TC, Wallenstein M (2007) Microbial stress-response physiology and its implications for ecosystem function. Ecology 88:1386–1394

Tanner K, Vilanova C, Porcar M (2017) Bioprospecting challenges in unusual environments. Microb Biotechnol 10:671

Westmann CA, Guazzaroni ME, Silva-Rocha R (2018) Engineering complexity in bacterial regulatory circuits for biotechnological applications. Msystems 3:2

Xiang S-R, Doyle A, Holden PA, Schimel JP (2008) Drying and rewetting effects on C and N mineralization and microbial activity in surface and subsurface California grassland soils. Soil Biol Biochem 40(9):2281–2289

12

Foresight: Methodologies, Potential Applications and Limitations

Cassidy Nelson, Ilker Adiguzel, Marie-Valentine Florin, Filippa Lentzos, Rickard Knutsson, Catherine Rhodes, Paul Rutten and Annika Vergin

Abstract Rapid developments in the fields of synthetic biology and biotechnology have caused shifts in the biological risk landscape and are key drivers of future threats. From a security perspective, extending our understanding beyond current risks to include emerging threats in these and related fields can play a vital role in informing risk mitigation activities. Insights that are generated can be combined with other efforts to identify vulnerabilities and prevent undesirable outcomes. Emerging risks that may occur at some point in the future are inherently difficult to assess, requiring a systematic approach to examining potential threats. Foresight is a process to consider possible future scenarios. Comprising a range of methods and

---

C. Nelson
Future of Humanity Institute, University of Oxford, Oxford, UK

I. Adiguzel
Engineer Research and Development Center, US Army Corps of Engineers, Vicksburg, MS, USA

M.-V. Florin (✉)
International Risk Governance Center, École Polytechnique Fédérale de Lausanne, Lausanne, Switzerland
e-mail: marie-valentine.florin@epfl.ch

F. Lentzos
Department of War Studies and Department of Global Health & Social Medicine, King's College London, London, UK

R. Knutsson
National Veterinary Institute (SVA), Uppsala, Sweden

C. Rhodes
Centre for the Study of Existential Risk, University of Cambridge, Cambridge, UK

Biosecurity Research Initiative at St Catharine's College, Cambridge, UK

P. Rutten
Interdisciplinary Biosciences DTP, University of Oxford, Oxford, UK

A. Vergin
Bundeswehr Office for Defence Planning, German Armed Forces, Berlin, Germany

techniques, foresight processes can offer novel insights into emerging synthetic biology and biotechnology threats.

This chapter offers an introduction to foresight, including definitions of key terms that could support a shared lexicon across NATO partners. An overview of different foresight methodologies, their potential applications, and different strengths and limitations are presented. As a key first step, an approach to selecting appropriate questions to guide foresight activities is suggested. Example questions for synthetic biology and biotechnology are highlighted. At the end of the chapter, the authors offer recommendations for the design of a foresight process, with the intention of providing a useable resource for NATO partners investigating emerging synthetic biology and biotechnology threats.

12.1 Introduction

The study of the future dates back to antiquity. Understanding what could lie ahead was of strategic importance to rulers and military leaders and was of great general interest to ancient societies and religions. While cultures developed different ways of thinking about the future that evolved over time, a historically common point of view was that there was one single predetermined future (Cuhls et al. 2012). The systematic study of different possible futures, and how these could be shaped by present actions, emerged as a new field of inquiry in the mid-twentieth century, in part due to pioneering work conducted by the Research and Development (RAND) corporation in Santa Monica, California (Kaplan et al. 1950; Helmer 1967). Since its emergence, the field of futures studies has undergone rapid expansion with the refinement of its conceptual underpinnings and development of different methodologies.

Today, futures work is undertaken by governments, militaries and scientific institutions, and other interested groups, with the aim of gaining actionable insight into possible emerging futures. In fields like synthetic biology and biotechnology that are undergoing rapid and continuous change, the ability to gain strategic insight from possible futures is highly relevant to policy development, risk assessment and threat analysis. It is particularly important to be able to identify the underlying drivers, range of uncertainty, points of convergence, and potential opportunities and challenges in these developing fields, and how these might be affected by particular policy interventions. For all of these, foresight – a process of conducting futures work – can offer strategic insight.

Foresight has been defined in multiple ways. In this chapter, we use an understanding commonly found in the literature, which highlights that it is first and foremost a process that involves "systematically attempting to look into the longer-term future of science, technology, the economy and society" through which "one comes to a fuller understanding of the forces shaping the long-term future" (Martin 1995; Miles 2010). Foresight, therefore, differs from forecasting, in that it does not aim to predict the future. Although the two terms are sometimes used interchangeably, forecasting is concerned with making "a probabilistic statement, on a relatively high confidence level, about the future" (Martin 2010).

Foresight can also be contrasted with hindsight, which is a systematic examination of the past. While the past offers useful information that can inform a foresight process, hindsight has access to outcome information that foresight does not. Care should be taken when combining these processes in order to avoid "observation selection effects" or biasing thinking towards historical occurrences (Fischhoff 1975).

Many additional terms specific to futures studies have been introduced and refined in the literature, with some confusion arising given shared and contradictory wording used in colloquial contexts (Trump et al. 2019). In order to use clear terminology in describing futures research, with the aim of developing a consistent lexicon across NATO partners, a definitions list is provided below. This is followed by examples of foresight research conducted on the topics of emerging synthetic biology and biotechnology.

The two subsequent sections provide an overview of different foresight methodologies and present an approach to foresight question choice, highlighting some specific questions for synthetic biology and biotechnology. Finally, recommendations for the design of a foresight process are offered, with the intention of providing a useable resource for NATO partners investigating emerging synthetic biology and biotechnology threats.

12.1.1 Foresight Terminology

Audience The individuals or organisations that are the intended recipient or end-user of the foresight process output. This might include, for example, government, business, military, civil society or broader public groups.

Backcasting Starting from a point in the future, analysing backwards in time the steps required for that future to occur.

Brainstorming A process in conversations or workshops used to develop a list of issues, drivers or ideas on a topic.

Cross-Impact Analysis An exploratory method to investigate the positive and negative effects between different interacting outcomes. It assumes that future events do not occur independently from each other, but rather the development of one influences the development of another.

Delphi Method A structured method used to gather and systematically prioritise expert views on the future.

Driver A key force or trend that is likely to have an impact in a relevant area.

Fifth Scenario A method to add a new scenario to an already existing set of scenarios to improve the understanding of the impact of the factors and drivers.

Futures Studies Structured approaches used to explore possible futures.

Forecasting A process used to make predictions about the future, commonly involving the use of past data and analysis of trends.

Foresight A process which systematically attempts to examine possible futures and determine the drivers shaping the longer-term future.

Horizon Scanning A specific foresight technique used during investigation which serves as a basis for analysis.

Participants The people taking part in a foresight exercise.

Roadmapping A method to showcase how a range of inputs, such as certain technology trends or policy changes, may combine in future development of the area of interest.

Red Teaming A process by which an external and independent group challenges an organization by assuming an adversarial role, with the aim of identifying weaknesses and gaps that can be addressed.

Scenario Analysis A process by which a range of identified possible future events are examined by considering alternative possible outcomes.

Scenario Creation Building scenarios for the purpose of a foresight exercise. It can be subdivided into four or more phases, including: *Investigation* (e.g. Horizon Scanning), *Analysis* (e.g. Driver identification; Uncertainty Analysis; Cross Impact Analysis), *Projection* (e.g. Scenario Writing), *Implications* (e.g. Backcasting; SWOT analysis), *Communication* (e.g. Trend reporting) and *Monitoring* (e.g. Trend monitoring).

Scenario Construction A generic term that summarizes various methods for scenario creation, with the end result being generation of generic scenarios which are transferred into narrative format for ease of communication.

Scenario Writing A method to create coherent scenario storylines in order to communicate them clearly to an audience.

SWOT Analysis A method of identifying the Strengths, Weaknesses, Opportunities and Threats (SWOT) in an area of interest.

Trend A general course, prevailing tendency or emerging pattern that suggests a change or particular trajectory.

Trend Analysis A process by which past data is examined and a pattern discerned.

Uncertainty Analysis A method used to identify key factors through analysis of possible impacts and their probability.

Wild Card An event that has a very low probability of occurring, but would have very high consequences.

12.1.2 Examples of Foresight for Biological Threats

While futures research is relevant to many fields, this chapter will focus on its applications to emerging synthetic biology and biotechnology threats. Conducting foresight in these domains is highly relevant to governments, military institutions, and a range of industry, academic and civil society groups given the rapid speed at which developments are taking place. There are a growing number of ways in which synthetic biology and biotechnology may pave the way to novel and high consequence risks while simultaneously offering new opportunities to address them (Hauptman and Sharan 2013). Foresight processes can help to avoid technological surprise and unexpected societal impacts, in part through identifying possible security threats before they emerge. When done well, insights from foresight exercises in synthetic biology and biotechnology can be used to inform action to avoid undesirable futures.

Foresight research for emerging synthetic biology and biotechnology issues has already taken place in a variety of settings. The following section highlights some recent published examples.

Using a scenario analysis 'causes and consequences' method, a 2013 study investigated reasons why an international Genetically Engineered Machine (iGEM) competition might hypothetically be closed down (Frow and Calvert 2013). The group quickly identified biosecurity and biosafety concerns, including synthetic biology competitors engineering pathogens with pandemic potential, as a possible reason. In another horizon scanning study looking at environmental threats to the UK, emerging biotechnology methods to genetically engineer pathogens were found to pose a 'high' risk of unintended consequences for biodiversity if released (Sutherland et al. 2008).

In one 2013 foresight study on emerging technologies, new gene transfer technologies and synthetically engineered biological agents ranked amongst the top ten risks, when prioritised by threat intensity and potential for misuse (Hauptman and Sharan 2013). The authors highlighted that security policy can be informed by adopting a long-range perspective where awareness would enable mitigation of threats that might otherwise be unaddressed.

A 2015 Delphi study asked 63 experts from government, academia, industry and non-governmental organisations how they perceive the bioweapons threat (Boddie et al. 2015). It found a wide diversity of opinion, including on the likelihood of an attack in the next 10 years with significant difference between biological scientists and other participants. Use of biological weapons by nonstate actors, especially religious extremists, or covert use of biological weapons by a state, were seen as much more likely than overt use by state actors. However, in this study, the diversity of participant views led the authors to conclude that assessing risks that research would be misused will be challenging: "a red line for what is allowable and what is forbidden in the name of security may not be clearly defined, and the way forward will be nuanced and complicated, possibly requiring a case-by-case evaluation with guidelines agreed upon by the scientific and policy communities" (Boddie et al. 2015).

In a 2017 report, scenario development was used to assess the changing security landscape in light of new genome editing technologies (Kirkpatrick et al. 2018). In this study, plausible future scenarios were developed through a series of workshops involving subject matter experts and published research. Using this approach, the authors were able to examine the current vulnerabilities and risks and couple each scenario with policy options to address governance gaps identified.

A 2017 study used a modified Delphi technique to examine emerging issues in biological engineering (Wintle et al. 2017). Twenty-seven experts from a diverse range of backgrounds participated in a horizon scanning exercise and identified 70 potential issues. Through voting and a workshop to systematically discuss the shortlist, 20 top-scoring issues emerged. These included issues highly relevant to synthetic biology and biotechnology threats, including synthetic gene drive developments, accelerated defence agency research in biological engineering, including the Defense Advanced Research Projects Agency (DARPA) Insect Allies Program, and the emergence of robotic 'cloud labs.'

In 2018, a consensus study report by the National Academy of Sciences in the US proposed a framework for assessing concern associated with a new technology in the field of synthetic biology (National Academies of Sciences, Engineering, and Medicine 2018). The framework was comprised of four main components: the technology's usability, the potential for use as a weapon, the requirements of actors and the potential for mitigation. Combined, this allowed for the ranking of synthetic biology-enabled capabilities by level of relative concern, with the highest being: recreation of known pathogenic viruses, biochemical production via *in situ* synthesis, and modifying bacteria to be more transmissible and/or lethal.

The above research highlights the role that foresight studies can serve in identifying novel threats in synthetic biology and biotechnology. The value in conducting formal, structured exercises comes from collating ideas and perspectives from a range of participants. Bringing together experts from diverse fields and backgrounds for a foresight study can enable identification of convergence points of technologies, which is highly relevant to synthetic biology and biotechnology threats. The following section will discuss in more detail different foresight methods, and highlight their strengths and limitations.

12.2 Foresight Methods

Since foresight emerged as an approach to futures studies, multiple techniques have been developed. Broadly, these can be divided into qualitative and quantitative methods (Fig. 12.1). It should be noted that often multiple methods are combined to examine a foresight research question. A general approach to foresight is presented below, followed by a non-exhaustive examination of common foresight methodologies, outlining their intended use, how they are applied, and their inherent strengths and limitations.

12.2.1 General Approach

While foresight methods vary widely, the majority share a common approach involving taking a range of inputs – such as historical data, published literature, technological trends or expert opinion – and conducting a structured exercise involving analysis, interpretation and prospection (Voros 2003). The type of foresight process selected will inform the tools that can be employed at this stage, which can include trend analysis, driver identification, and envisioning possible future scenarios. The output of a foresight exercise can be tangible or intangible, and might include identification of the range of next-step options, or general changes in thinking about the topic under scrutiny. Foresight processes should be set up so that the outputs can support action, for example by being able to directly inform planning, policy and strategy.

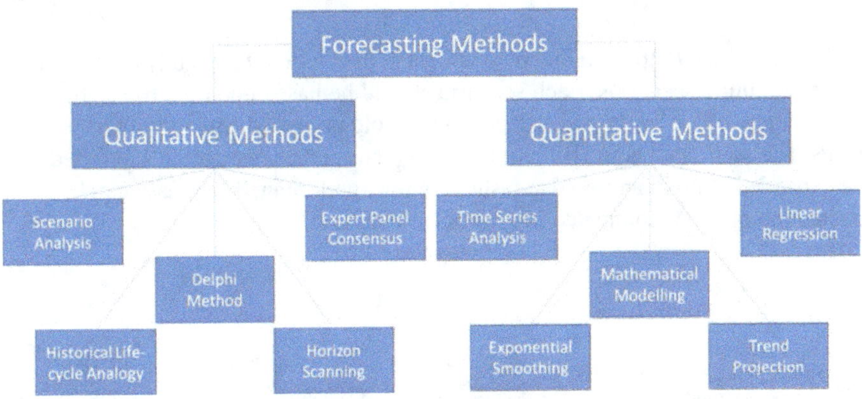

Fig. 12.1 Overview of some common forecasting methods

12.2.2 Delphi Method

One of the first foresight techniques developed, the Delphi Method is still widely used today.[1] This method employs a structured group communication process to allow future scenarios to be constructed. Using a multi-step approach, the Delphi Method allows for individual expert contributions and feedback with the aim of reaching participant consensus on all posted responses (Linstone and Turoff 1975). Carrying out a Delphi exercise requires planning and preparation, with the whole process taking weeks to months. The phases are outlined below.

Phase A: Participant Selection

The first phase consists of selecting and inviting individuals to participate in the foresight study. This phase is important because the outcome of the Delphi exercise will depend upon the knowledge and range of expertise of the participants. It is recommended to aim for between 15-35 specialist participants (Renzi and Freitas 2015). Invitations should be sent to experts from a range of backgrounds, which might include technologists and representatives from civil society, the natural and social sciences, humanities, and the defence, intelligence and security services. Achieving diverse participation should bring the relevant expertise without biasing the outcome to a narrow perspective from one domain. Participant selection should be guided by the foresight research question and consideration should be given to ensuring cultural and demographic diversity.

Phase B: Questionnaire

After participants have accepted the invitation to take part in the Delphi exercise, a strategically formulated questionnaire should be sent to each individual. Questions should be clearly phrased and their objective transparent, but they should not be leading. Answers should be collated anonymously into a single document which is sent to all participants for review and comment, with the aim of reaching a consensus on its contents.

Phase C: Future Scenarios

Following *Phase B*, the combined document is then used by the study organisers to construct future scenarios. Each scenario should be based upon the foresight exercise theme, and the total number generated should be agreed with the expert participants. The scenarios should be built from the convergence of material submitted and be expanded upon in an iterative fashion with expert input. This phase can be conducted remotely, or completed in a workshop format.

---

[1] The Delphi Method was named after the Greek Oracle of Delphi, the high priestess of the Temple of Apollo who was said to be able to prophesize the future, although this foresight method's reliance on collated expert opinion and group consensus to deliver multiple possible scenarios makes the analogy imperfect.

Phase D: Result Analysis

The resulting scenarios should be analysed to understand their implications. In particular, actions that can be taken to steer towards desirable future scenarios or away from identified risks should be summarised explicitly. The results of scenario exercises can be condensed into a report with the intended target audience in mind.

12.2.2.1 Strengths

The Delphi Method has a variety of strengths depending on how the exercise is prepared. Bringing together a range of subject matter experts from diverse but relevant fields allows for the generation of multiple ideas that cover different perspectives. Because participant opinions are initially solicited individually, groupthink can mostly be avoided. A consensus can be reached rapidly, and the ideas generated can be refined by experts in the later phases. Some of these viewpoints would be difficult to extract from the literature or other sources, and therefore this method is useful to generate ideas in an area where data is lacking. The Delphi Method is also attractive because it is relatively low cost to administer, especially when conducted remotely via email.

12.2.2.2 Limitations

This foresight technique is ultimately bound by how it is initially constructed. Poor selection of participants, questionnaire content, or route to later expert input, can limit the usefulness of its results. Because consensus is required in Phases B and C, the Delphi Method struggles when there is a wide range of differing opinions or conflicting views. The method is also prone to various biases, being constrained by the knowledge and opinions of the participants, and carrying the risk that the facilitator's viewpoints and interpretation can dominate the analysis in the final report. The Delphi Method can also be quite time consuming, both for those leading it – with each phase lasting several weeks to allow expert input to be gathered – and for participants whose active involvement is needed throughout the process.

12.2.3 Horizon Scanning

There is ambiguity around the meaning of the term "horizon scanning" as it is often used interchangeably with "foresight" and "futures." However, although some variation exists, amongst practitioners in the field horizon scanning is generally a well-defined foresight method. A standardized approach to horizon scanning is outlined below.

The UK Government Office for Science Futures Toolkit presents a 'Three Horizon Model' in which Horizon 1 issues fall within current policy and strategy, Horizon 2 issues are those that increase in importance in the medium term, and

Horizon 3 (H3) issues are new challenges and opportunities that have not yet begun to emerge, and whose drivers are difficult to spot (Government Office for Science 2017). Horizon scanning activities are generally aimed at H3: There is no commonly accepted timeframe for this horizon; in some contexts horizon scanning work may look 5 years ahead and in others a 20 year gap may be considered.

The aim of horizon scanning is to identify and understand so-called "weak signals". These are events, trends and developments taking place today that could shape H3. In most cases, it will not yet be clear what their impact will be on H3, which are the more and less important signals, and how they may evolve or converge in the future.

Horizon scanning involves a group of participants who collect information and opinions, usually over the course of at least several weeks. Some horizon scanning exercises ask participants to perform "scans" at regular intervals (e.g. once a week). These scans are brief, often no more than one-page reports on possible signals to be considered. In a very open-ended exercise, participants are given the freedom to report on any signal they feel is relevant, based on any source of information. In a more constrained horizon scanning exercise, individual participants may be asked to consider particular issues, or monitor a specific source of information (e.g. a particular academic journal).

At the end of the exercise, all scans from every participant are compiled into a single document which synthesizes their findings and highlights recurring themes. Other approaches are possible; some horizon scanning activities adopt a phased structure, involving individual submissions first, and later collective refinement and prioritising of issues. In many organizations, the results of horizon scanning are used to inform further futures work, such as scenario exercises or a Delphi Method activity (see above).

12.2.3.1 Strengths

Horizon scanning exercises are relatively easy to organise, and can be performed without a physical meeting taking place between participants. They also enable a diverse group of stakeholders to be engaged in considering the future. Participants may range across disciplines and backgrounds, bringing together a variety of perspectives.

Another strength of horizon scanning is its emphasis on external stakeholders and sources; the technique is not reliant on in-house expertise or knowledge. Rather, the aim is to filter through a large volume of data, leveraging many stakeholders to distill a list of weak signals or identified issues that could influence H3. Horizon scanning is inherently inclusive; its aim is not to narrow down this list, so a large number of signals are usually considered and included. Subsequent phases may include work to prioritise or rank signals or issues.

12.2.3.2 Limitations

Although offering many advantages, the very wide net thrown by horizon scanning exercises is an important limitation to consider. Horizon scanning usually produces a large list of weak signals, but many of the signals identified may be irrelevant or unimportant. This can be mitigated by using a phased approach, or subsequently trimming the list with other futures techniques, e.g. to produce a set of scenarios for further expert elicitation.

Horizon scanning also relies to some degree on human intuition. In many cases, even for signals which will ultimately prove critical, there is little evidence available to assess their importance. It is important that those taking part in a horizon scanning exercise, and those who will be using its output, understand and accept this limitation. It is inherent in the long-term focus of horizon scanning that many of its results will not be robustly evidence-based.

12.2.4 Scenario Analysis

Scenario Analysis is a useful method to identify the multiple ways situations in the future might evolve. This technique can help decision makers develop plans to exploit opportunities, reduce uncertainties and manage risks the future may hold. Additionally, the monitoring of indicators embedded in various scenarios can create early warning signals of likely future trajectories. Scenario analysis is recommended when a situation is complex or single predictions are too uncertain to trust.

If used in the initial stages of national policy formulation or long-term corporate strategies, scenario analysis can have a strong impact on decision making. This method provides a set of plausible and possible futures for which decision makers may need to consider. It is also useful as a tool for strategic planning processes that brings together decision makers or stakeholders with analytical experts to work on alternative futures for which they must plan. Engaging stakeholders and decision makers in the scenario analysis process can generate commitment for the projects, save time and produces more useable results.

Scenario creation, which surrounds the process of scenario analysis, can be subdivided into different phases, including: *Investigation* (e.g. Horizon Scanning), *Analysis* (e.g. Driver identification; Uncertainty Analysis; Cross Impact Analysis), *Projection* (e.g. Scenario Writing), *Implications* (e.g. Backcasting; SWOT analysis), *Communication* (e.g. Trend reporting) and *Monitoring* (e.g. Trend monitoring). In order to create scenarios including implications at least the first four phases are required.

There are different techniques for generating scenarios, e.g.: *Simple Scenarios*, *Cone of Plausibility*, *Alternative Future Scenarios*, *Explorative Scenario Construction* with or without quantitative analysis. These methods have been referred to by various names in the literature.

12.2.4.1 Strengths

The use of scenarios offers the possibility to describe many different possible and plausible futures. Comparing multiple scenarios in the analysis makes it easy to identify factors or drivers that are essential for future developments, whether desired or not. In addition, scenario analysis can be used to test assumptions about the future or even find and warn against critical developments.

A very important advantage of scenario analysis is the possible involvement of decision-makers and stakeholders in the scenario process. This promotes a high level of understanding for the various possible future developments and achieves a high commitment to the actual work.

12.2.4.2 Limitations

The results of a scenario analysis exercise can be misleading if the group of participants are too homogeneous with limited diversity, falling prey to groupthink. For the method to be successful, it is necessary to have creative thinking and prospection far into the future about a variety of plausible possibilities. If the participants in the scenario analysis do not fully engage with this step, this will severely affect the quality of the results. In addition, it is important that the analyst for the exercise is an expert in the analytic techniques and also has a deep understanding of the subject matter. This is critical to ensuring appropriate analysis of the quality of the key driving factors and the assumptions that are present throughout the exercise.

12.3 Foresight Questions

The formulation and selection of question(s) are very important components of the foresight process. Compared to approaching a topic broadly, well-constructed questions enable a tailored exercise. Not only should the question encapsulate the subject matter to be examined, but it should also lead towards actionable content in the final output. The following section covers an approach to question choice and lists specific questions that could be considered for emerging synthetic biology and biotechnology threats.

12.3.1 Approach to Choosing Questions

For any given area of interest, there could be a range of questions that a group would like to answer about possible futures. Framing the question(s) carefully at the beginning of the foresight process will help define the study's aim and scope, and have an important impact on the quality of the process. Questions should be designed such that they are:

- Framed in broad, rather than specific, terms
- Open and not too narrowly focused
- Stimulating discussion and thinking instead of leading to a particular endpoint
- Within a specific timeframe (e.g. in the next 15 years)

Participants in a foresight exercise should be made aware of the overarching question as early as possible in the process. The question itself need not overly constrain the final report or its conclusions; unanticipated areas of interest might arise during the process, for example. At the same time, questions should be formulated so that they avoid out-of-scope deviations that might hinder the topic of interest being fully addressed.

12.3.2 Questions for Synthetic Biology Threats

Concern about potential emerging threats associated with synthetic biology and biotechnology and related fields can lead to a range of future-focused questions of strategic interest to national governments, military leaders, and a range of other groups. The focus of these questions could, for example, range from future risks associated with a particular form of biotechnology or a new synthetic biology development, advancements more broadly in these fields, or on convergence of two or more trends within these and related fields.

Often it will not only be emerging threats which are of interest when seeking to apply foresight to synthetic biology and biotechnology. There is also great interest in more positive aspects of the transformative potential of these technologies and the ways in which developments might also help to address security threats. While some questions and the associated processes may focus only on one aspect, there will also be value to foresight processes which combine these. In addition to this, it is not only technological developments that will influence the future security impacts of synthetic biology and biotechnology, but also developments in economic, social and political contexts, and changes to the policy and regulatory environments. Questions can be designed that will incorporate some of these developments as well.

Some example questions in this area, formulated during the workshop, that may be of strategic interest include:

- Over a 15 year timeframe, what are the potential impacts of synthetic biology developments converging with other disruptive technologies?
- In the next 10 years, what regulatory and legislative gaps will be prominent if developments in gene-editing technology converge with a widespread availability of low-cost desktop DNA synthesizers, enabling the capability of practically any lab to design, engineer and print pathogen genomes?

- If, in the next 5 years, a do-it-yourself synthetic biologist were to accidentally or deliberately release a contagious human pathogen, what impact would this have on the research community as a whole and Code of Conduct creation?
- Will current information controls, such as embargoes on publication of dual-use synthetic biology research, be suitable in 20 years?
- In the next 10 years, how will emerging DNA synthesis technology impact screening for potential synthetic biology threats?
- How will developments in delivery mechanisms of biological agents affect synthetic biology threats over the next 5 years?

12.4 Recommendations

The use of foresight techniques to generate insightful and actionable information about emerging synthetic biology and biotechnology threats has great potential. To increase the effectiveness of foresight, careful consideration should go into process design, including the choice of methodological approach, questions and participants. Potential constraints on the resources and capacities of the organisers and participants should be taken into account at this stage as well.

The intended audiences and users should be aware of the limitations and inherent uncertainties of foresight in general, as well as the particular process used. In general, the further out in time a foresight process is being used for, the more speculative the results will be, with less resolution and a greater chance of unanticipated and wildcard events in the interim. It is particularly important for users to remain cognizant of the difference between foresight and forecasting; foresight does not aim to predict the future. Below, some recommendations on the foresight process design are provided, followed by specific considerations for emerging synthetic biology and biotechnology threats.

12.4.1 Foresight Process Design

Careful consideration should go into the preparation and design of a foresight process. A deliberately constructed exercise will be more likely to deliver a relevant and actionable outcome for the intended end-users. The process should be designed keeping in mind the available resources and necessary timeframe. The target audience should be a key consideration and the foresight methodology should be made transparent to them with the limitations explicitly acknowledged.

Answering the following five questions will help in the foresight process design, and a worked example based on a synthetic biology foresight question, which was examined during the Lausanne workshop, is provided for illustration.

1. For what specific purpose is foresight being done?

2. What are the objectives?
3. Who is the audience?
4. What are their expectations?
5. What process is required?

Example Foresight Question Over a 15 year timeframe, what are the potential impacts of synthetic biology developments converging with other disruptive technologies?

1. *For what specific purpose is foresight being done?*
   (a) In order to avoid technological and societal surprise.
2. *What are the objectives?*
   (a) To engage and inform key stakeholders;
   (b) To enable the development of appropriate governance mechanisms, for example through informed regulatory export control;
   (c) To gain insight into adversarial capability;
   (d) To drive the creation of new capabilities and realize emerging opportunities;
   (e) To identify key trends and early warning signals;
   (f) To understand and identify threats;
   (g) To avoid technological surprise;
   (h) To create conceptual connections to inform strategic action.
3. *Who is the audience?*
   (a) Civilian population
   (b) Military leaders and Defense Department
   (c) Policymakers
   (d) Funders
   (e) First responders
   (f) Non-Governmental Organizations
   (g) Intergovernmental Organizations (UN, WHO, OIE)
   (h) Academic researchers
   (i) Intelligence and security services
4. *What are their expectations?*
   (a) To generate specific, actionable recommendations;
   (b) To reduce uncertainty while also assessing and indicating the degree of uncertainty;
   (c) Raise awareness and engagement with possible futures;
   (d) Identify key possible future events, including their associated forks, nodes and branch points;
   (e) Identify possible interventions;
   (f) Encourage dialogue between stakeholders;

   (g) Note: expectations may need to be managed, for example, explicit communication about foresight not being a tool to generate accurate predictions about the future.

5. *What process is required?*
   (a) Delphi Method with horizon scanning component;
   (b) Identification of the key drivers and scenario analysis;
   (c) SWOT analysis;
   (d) Sharing of findings with the target audience.

12.4.2 Synthetic Biology Considerations

When designing a foresight exercise specifically for emerging synthetic biology and biotechnology threats, due consideration should be given to a number of factors. The participants involved in the exercise should be chosen for their expertise, but not limited to a certain discipline: the value in a foresight process is achieved through harnessing opinions from across domains, and even for technology-specific foresight questions, fields such as the social sciences and humanities and the security community have important perspectives to add. In addition, the inclusion, from the design phase onwards, of foresight expertise will have great value for the proper execution of the process.

Emerging synthetic biology and biotechnology inherently have sensitivities that require consideration. For example, certain developments may be occurring within confidential industrial contexts, and concerns about dual-use applications may prevent certain scientific findings from being openly discussed. In addition, security concerns and evolving threats are likely to be confidential in nature. All of these may skew the inputs going into a foresight exercise. However, this need not undermine the value of a well-designed process, that does not breach security concerns, for strategic insight. Foresight exercises at different confidentiality levels could be considered in the military context and combined for use by decision-makers.

For many fields, but for synthetic biology in particular, emerging threats are likely to arise at the convergence points of new developments. This highlights the importance of ensuring foresight exercises are designed to allow for broad thinking on how scientific knowledge and technology can enable threats and reduce the risk threshold level.

Foresight can have a significant impact on strategic decision-making and direct work within a wide range of military contexts and government institutions, but it is important to acknowledge that evaluation of a foresight exercise's utility is inherently difficult. Iterative processes that allow for the systematic re-examination of the outputs of a foresight exercise, both at regular intervals and as significant or unpredicted wildcard events take place, will increase the likelihood of foresight offering substantial benefits.

12.5 Conclusion

The ability to consider future scenarios in a systematic fashion is highly relevant to examining emerging synthetic biology and biotechnology threats. From both a governmental and military perspective, extending understanding beyond current risks and being able to take proactive steps in addressing vulnerabilities. Mitigating emerging threats is preferable to being surprised by and reactive to technological developments. Insights gained also have value for a range of other groups and stakeholders. Foresight offers a process by which to consider and explore possible future scenarios, and while any examination of the future has inherent limitations, if designed correctly, foresight exercises can provide strategically useful information for action.

References

Boddie C, Watson M, Ackerman G, Gronvall GK (2015) Assessing the bioweapons threat. Science 349(6250):792–793

Cuhls K, Blind K, Grupp H (2012) Innovations for our future: Delphi'98: new foresight on science and technology, vol 13. Springer

Fischhoff B (1975) Hindsight is not equal to foresight: the effect of outcome knowledge on judgment under uncertainty. J Exp Psychol Hum Percept Perform 1(3):288

Frow E, Calvert J (2013) Opening up the future (s) of synthetic biology. Futures 48:32–43

Government Office for Science (2017) Futures toolkit for policy-makers and analysts. UK Government: 1.0. 2017 Nov. Available: [Internet]. Available from: https://assets.publishing.service.gov.uk/government/uploads/system/uploads/attachment_data/file/674209/futures-toolkit-edition-1.pdf

Hauptman A, Sharan Y (2013) Foresight of evolving security threats posed by emerging technologies. Foresight 15(5):375–391

Helmer O (1967) Analysis of the future: the Delphi method. RAND Corporation Santa Monica California

Kaplan A, Skogstad AL, Girshick MA (1950) The prediction of social and technological events. Public Opin Q 14(1):93–110

Kirkpatrick J, Koblentz GD, Palmer MJ, Perello E, Relman DA, Denton SW (2018) Editing biosecurity: needs and strategies for governing genome editing. George Mason University

Linstone HA, Turoff M (1975) The Delphi method. Addison-Wesley, Reading

Martin BR (1995) Foresight in science and technology. Tech Anal Strat Manag 7(2):139–168

Martin BR (2010) The origins of the concept of 'foresight'in science and technology: an insider's perspective. Technol Forecast Soc Chang 77(9):1438–1447

Miles I (2010) The development of technology foresight: a review. Technol Forecast Soc Chang 77(9):1448–1456

National Academies of Sciences, Engineering, and Medicine (2018) Biodefense in the age of synthetic biology. National Academies Press, Washington, DC

Renzi AB, Freitas S (2015) The Delphi method for future scenarios construction. Proc Manufact 3:5785–5791

Sutherland WJ, Bailey MJ, Bainbridge IP, Brereton T, Dick JTA, Drewitt J et al (2008) Future novel threats and opportunities facing UK biodiversity identified by horizon scanning. J Appl Ecol 45(3):821–833

Trump BD, Cegan J, Wells E, Poinsatte-Jones K, Rycroft T, Warner C, ... Linkov I (2019) Co-evolution of physical and social sciences in synthetic biology. Critical Review Biotechnol 39(3):351–365

Voros J (2003) A generic foresight process framework. Foresight 5(3):10–21

Wintle BC, Boehm CR, Rhodes C, Molloy JC, Millett P, Adam L et al (2017) Point of view: a transatlantic perspective on 20 emerging issues in biological engineering. eLife 6:e30247

13

Potential Biosecurity Threats and their Prediction

Kaitlin M. Volk and Trajan J. Gering

Understanding what kinds of biological weapons can be made with what sort of technology and by whom is an important component of biosecurity. An equally important component is understanding the different potential targets a biological weapon could be designed to attack, how the weapon would be deployed against these targets, and the available strategies to detect the creation and deployment of an illegal biological weapon. Understanding how a weapon could be deployed against specific targets affords decision makers a better picture of the current state and capabilities of biowarfare and bioterrorism that need to be protected against. Understanding current detection capabilities allows for a more informed discussion on biosecurity tools, and, more importantly, allows for the identification of critical gaps and research needs to improve risk-screening, detection, environmental remediation, and various other normatively beneficial and legitimate uses of emerging biotechnologies (Trump et al. 2020a). These two topic areas are discussed in this chapter.

13.1 Methods of Deployment

The Center for Disease Control and Prevention (CDC) classifies disease transmission into two primary categories: direct transmission and indirect transmission (CDC 2011). Concerning direct transmission, individuals pick up the disease through direct contact with an infected person or from direct contact with an environmental reservoir. This also includes droplet spread, in which the close-range spray of a sneeze or cough transmits a pathogen without the assistance of an intermediate object. Examples of natural outbreaks spread primarily through droplets

---

K. M. Volk (✉) · T. J. Gering
ERDC-EL, US Army Corps of Engineers, Vicksburg, MS, USA

include the COVID-19 virus and the plague (Carniel 2002; Linkov et al. 2021). Additionally, researchers in 2012 modified a strain of avian influenza to be transmissible between mammals via respiratory droplets (Evans and Selgelid 2015). While direct transmission with droplets would restrict spread to only close-range encounters, it has proven to be a highly effective mode of transmission particularly in densely populated areas like cities.

In indirect transmission, a pathogen from a sick individual is left on/in an intermediate object and is transmitted to a healthy individual when they come into contact with the infected object. The CDC delineates three categories of indirect transmission: vehicle, vector, and airborne. In vehicle-based transmission, inanimate objects like food, water, biological products (e.g. blood), and fomites (e.g., clothing and utensils) are contaminated with the pathogen. A number of bioterrorism acts that utilized vehicle transmission exist. Letters were used in the 2001 anthrax attack that targeted media outlets and politicians. The Rajneeshee cult poisoned salad bars with *Salmonella* in order to influence a local election, using food as the vector for their biological attack. Researchers modeling a bioterror attack determined that terrorists could affect several hundred thousand individuals if they could cultivate and deploy enough biotoxin into a single milk-processing facility (Wein and Liu 2005). This is because the dairy processing industry is so heavily centralized. Many realms of the agriculture and food production industry are similarly centralized, increasing the risk of the transmission of pathogens to consumers through the ingestion of contaminated food or drink (Sobel et al. 2002). Contaminated blood, medicine, and other inanimate objects could also be used as vehicles (Shinwari et al. 2014), and the vehicle method could be particularly effective at attacking sequencing machines, as discussed further below (Faezi et al. 2019).

In vector-based transmission, organisms carry a pathogen and transmit it upon contact with humans (CDC 2011). Animal vectors are a common mode of transmission of natural outbreaks (e.g. mosquitos transmitting malaria and rats transporting the plague) (CDC 2011; Carniel 2002) and have been used in previous biological attacks (e.g., plague-ridden fleas released into China by Japan) (Franconi et al. 2018; Zilinskas 2017). Additionally, the majority of the novel human infectious agents that have emerged over recent decades have been zoonotic, meaning they were transmitted from animals to humans following a mutation (Franconi et al. 2018). Despite their proven effectiveness, little attention has been paid to animals as a means to start a biological attack, potentially due to the difficulty of obtaining, keeping, and releasing large amounts of any organism discreetly.

While much of the published biosecurity literature does not address specific modes of pathogen transmission, the airborne mode has received the most attention for its potential use in a biological attack (Yeh et al. 2013, NRC 1997, Anand 2018). In airborne-based transmission, pathogens are attached to dust or droplet nuclei suspended in the air. Concerns over the aerosolization of pathogens have existed for decades. The US Army Chemical Corps ran a series of experiments in the 1950s and 1960s to understand patterns of aerosol dispersal of biological weapons over short distances in cities, medium distances in rural communities, and extremely long distances across entire continents (NRC 1997). Using planes, the Chemical Corps dispersed zinc cadmium sulfide, *Serratia marcescens,* and *Bacillus globigii* over 33 urban and rural sites and large swaths of land from ocean to ocean and border to border. These latter experiments are known as Operation LAC, standing for "Large Area Coverage," and demonstrated that airborne biological weapons could cover extremely large geographical areas (NRC 1997). Japan, Iran, and Iraq at different times created bombs or missiles designed to aerosolize pathogens upon explosion (Zilinskas 2017). Similar concerns exist today with malicious terrorist or cult groups (Zilinskas 2017; Shinwari et al. 2014), in addition to concerns over the possibility of these groups using drones to remotely disperse pathogens (Anand 2018) or concealed aerosol containers to release an airborne pathogen in crowded and confined areas (Yeh et al. 2013).

Overall, while a synthetic pathogen could be spread in a variety of ways, the airborne mode of transmission would likely be the most effective mode for a large-scale biological attack as it could be easily released by land or air, spread over a large area in a relatively short amount of time, and easily enter people's bodies through respiration.

13.2 Targets

Four main categories that biological weapons could be engineered to attack were identified in the literature. These targets include humans, agriculture, technology, and the environment. The ease with which these attacks could be carried out, consequences of the attack, and potential strategies to prevent the attack are presented below for each target.

13.2.1 Humans

Unsurprisingly, much of the focus on biological weapon production and defense have used humans as the target. A biological attack against humans could result in direct human deaths or illness, widespread fear and panic, economic loss, and costs related to the response and remediation of the pathogen (Elbers and Knutsson 2013). Human pathogens can be readily obtained from the environment or natural outbreak events and are found in laboratories of various security levels around the world (Gronvall 2015). Most of the work being done on synthesizing viruses from DNA ordered through DNA synthesizing organizations is done on human pathogens (De Vries 2017, Noyce et al. 2018, Berger 2019). With a few exceptions, including the U.S.'s use of biological weapons against plants in Vietnam, the majority of state-sponsored synthetic biological weapon programs have focused on zoonotic human pathogens, especially *Bacillus anthracis* and *Yersinia pestis* (Kelle et al. 2010; Zilinskas 2017).

The probability of an attack in the near future with biological weapons from state-actors, sub-state actors, and non-state actors seems to be admittedly low based on the low number of previous biological attacks, but the outcome of such an attack would be catastrophic and warrants preventative measures. Biological attacks will also become more likely as the cost and effort associated with synthetic biology and the creation of pathogens decreases. As demonstrated by natural outbreak events, pathogens have an innate ability to persist in a given location indefinitely, essentially remaining dormant in human or environmental reservoirs between outbreaks. Thus, an initial biological weapons attack could have long-term and cascading effects unseen in attacks with more-traditional weapons (Plianbangchang 2005).

Many laboratories that handle human pathogens have protective measures commensurate in level to the danger of the pathogens they are handling. Biosecurity levels range from 1, requiring the least amount of protective measures, to 4, requiring the most. These regulations are put in place to reduce the risk of a pathogen being accidently released or of unauthorized personnel gaining access to the pathogen. However, not all laboratories around the world handling dangerous pathogens properly follow biosafety regulations, or take measures to reduce biosafety challenges via a 'safety-by-design' approach (Trump et al. 2020c). Outreach and support, both monetarily and intellectually, may be needed between countries to ensure all biological laboratories are properly protected. Biological journals also need to take responsibility for screening papers so that information that could be easily used by nefarious actors to create biological weapons doesn't become readily available. This debate over the risk of publishing dual-use research and the role of academic journals is evident in the ongoing scrutiny of a publication detailing the synthesis of horsepox, a close relative to the smallpox virus (Noyce and Evans 2018; Yong 2018), and is yet to be resolved.

13.2.2 Agriculture

Biological attacks against agriculture and livestock, also known as agroterrorism, would result in large negative consequences despite receiving less attention in the literature. Agroterrorism is defined as the introduction of an animal or plant infectious disease to induce fear in people, threaten social stability, and cripple a nation's economy (Yeh et al. 2013). Animal and plant infectious diseases include both viruses, such as foot in mouth disease (FMD), rinderpest, and avian influenza; and bacteria, such as anthrax, brucellosis, and glanders (Yeh et al. 2012). While synthetic viruses and bacteria are of the most serious concern currently, agroterrorism could be susceptible to larger multicellular organisms as advancements in synthetic biology are made on that front. Traditional agricultural pests or diseases, such as the Russian wheat aphid (*Diuraphis noxia*) or ring rot (*Clavibacter machinanensis*), could be modified to be more resistant to pesticides or environmental conditions, thereby increasing their potential damage (Anand 2018; Getz and Dellaire 2018).

Agroterrorism lacks some of the cognitive and cultural shock of bioterrorism and has rarely been used in the past. One of the few examples of an offensive agroterrorism program is the Japanese Unit 100, which was dedicated to creating biological weapons for use against animals during WWII (Zilinskas 2017). Agroterrorism could be an easier and safer avenue of attack for the attacker (Elbers and Knutsson 2013; Anand 2018). Animal and plant infectious diseases are often abundant in the natural environment, making them relatively easy to obtain. Would-be attackers can handle the disease with little to no threat to their own lives or the lives of those people they care about. For example, while the use of a human pathogen during war has a very high risk of spreading back to the home nation, agricultural pathogens will be more contained and less likely to backfire especially if strains of a crop or livestock specific to the attacked location are targeted. Farms are often unsecured and geographically dispersed, making it easy for attackers to deploy a pathogen. Livestock and crops are kept in highly concentrated areas and livestock are frequently exposed to other herds or flocks during long-distance transport, effectively providing a built-in dispersal mechanism. Lastly, livestock often lack any resistance to pathogens that are prevalent in other countries, and veterinarians or agricultural officials are less likely to test for or recognize diseases that aren't endemic to their region (Elbers and Knutsson 2013).

Overall, agroterrorism with a synthetic pathogen is an easier route of attack for a terrorist organization than a traditional biological attack against humans. A biological attack against important crops or livestock could have devastating effects on the economy and social well-being of a country, as has been seen in natural agricultural outbreaks (e.g. the Irish potato famine in 1845 or the mass culling of pigs infected with African swine fever in Ukraine in 2012). The threat of a biological attack against agriculture could be reduced by preventing actors from acquiring agricultural pathogens, increasing security on farms, developing early detection systems (including the education of farmers and veterinarians on foreign agricultural pathogen identification) and stockpiling livestock vaccines (Yeh et al. 2013; Anand 2018). Restrictions on the use of agricultural pathogens in synthetic biology experiments could also be applied in a way similar to those on human pathogens.

13.2.3 Technology

Synthetic biology has also been shown to be effective at attacking technology (Berger 2019; Ney et al. 2017). DNA is a relatively stable platform capable of storing large amounts of information. Normally this information is biological, but scientists have begun to look at DNA as a way to store large data and datasets, including images, audio, and videos (Berger 2019). To demonstrate the capability of technology to store and recover data from DNA, researchers at Harvard University recently used CRISPR-based tools to encode a GIF of a galloping horse into the genomes of living bacteria (*Escherichia coli*) (Shipman et al. 2017). However, if one is able to store data and code in DNA, then one should also be able to store malicious code.

In the same year that scientists at Harvard were storing GIFS in *E. coli*, scientists at the University of Washington were encoding malware into a segment of DNA to successfully gain remote control of the computer that sequenced the malicious DNA (Ney et al. 2017). Many next-generation DNA sequencing (NGS) systems use a 2-bit DNA encoding scheme to read and store genetic information – A is coded as 00, C as 01, G as 10, and T as 11. The authors of the study encoded a straightforward and commonly available computer bug and then transferred the computer code into the corresponding nucleotide sequence. They then bought this sequence from a gene synthesis company that creates synthetic gene fragments. Upon receiving the malicious synthesized DNA fragment and running the fragment through a sequencer and downstream analysis program, the authors gained full control of the computer attached to the sequencer.

It is important to note that the authors deliberately introduced a vulnerability into the analysis program they attacked prior to sequencing the malicious DNA. However, the authors of this study analyzed a number of NGS analysis programs and found that they are 11 times more likely to use insecure programming functions than non-NGS control programs ($p = 0.027$, Ney et al. 2017). They concluded that NGS programs do not follow best software security practices and need to be updated so as to be more robust against attacks. The authors were limited in the kind of malware they could encode into DNA because of current restrictions in the length of DNA fragments that can be sequenced and requirements for sequence stability. Advancements in NGS systems to sequence longer segments of DNA will increase the length and complexity of malware that can be encoded in DNA, and advancements in synthesizing DNA could overcome current structural requirements and also allow for a wider variety of malware. Advancements in DNA sequencing and synthesizing are ongoing as researchers innovate in pursuit of beneficial synthetic biology applications.

With the encoding of malware into DNA, any laboratory or company that sequences DNA becomes vulnerable to an attack on their computer system. Attackers could target the human biobanks of companies like 23andMe or AncestryDNA to expose the genetic information of earth's population. This could allow attackers to synthesize a pathogen that is fatal to a small subset of people with the targeted genetic sequence but is benign to the general population (Faezi et al. 2019). Attackers could also target high-level laboratories that work with pathogens to crash their computers or gain access to confidential files. Malicious DNA could be sent directly to these sequencing services (e.g., a spit sample sent to an ancestry database) or introduced by spraying malicious DNA on surfaces that are likely to be swabbed for DNA analysis (e.g. a countertop during a health inspection or criminal investigation) or come into contact with the sequencing machine (e.g. rubber gloves or lab coats) (Faezi et al. 2019).

As advancements in NGS systems are made, updates to bring NGS software in line with best security practices should be prioritized to reduce the risk posed by malicious DNA (Ney et al. 2017). Sequencing companies should also take more responsibility in identifying suspicious DNA sequences and denying or reporting the order. One group of researchers had already started to develop a technique to detect and exclude malicious DNA by checking the sequenced DNA against known sequences before it was fully processed (Islam et al. 2019). As this is a new emerging threat, more effort needs to be applied to increase security at every level, including synthesizing DNA, sequencing DNA, analyzing sequenced data, and securing valuable genetic databases.

13.2.4 The Environment

To date, no literature that discussed using synthetic biology to directly attack the natural environment has been found. There is ongoing research focused on using gene drives and horizontal gene transfer in the open environment for beneficial purposes, including modifying mosquitos so that they are resistant to malaria (Gantz et al. 2015) and modifying coral symbionts to be more resistant to coral bleaching (Cleves et al. 2018). There are inherent risks of releasing engineered organisms into the environment since the effects of such organisms on the local community are largely unknown. However, that debate has been ongoing for decades and is outside of the scope of biosecurity. Theoretically, it could become possible for malicious actors to utilize gene drives to purposefully collapse species critical to the functioning of an ecosystem without detection. However, the technique is complicated and still in its infancy. Gene drives would be particularly ineffective against long-living species with long generation times since it would take an exorbitantly long time to propagate the deleterious gene throughout the population through natural reproduction. Overall, biological attacks against the natural environment are a very indirect route of causing harm to an enemy. Humans and agriculture are much more likely targets since harm can be inflicted almost immediately and research in engineering these sorts of pathogens is more robust. Advancements in synthetic biology that enhance the capabilities of gene drives or other methods that could be deployed against the natural environment should still be tracked as this could become a target in the future.

13.3 Screening Techniques

Effective biosecurity should be able to identify when (a) a biological weapon is being produced in order to prevent its full production and deployment and (b) a biological weapon has been deployed in order to attribute the attack and to start remediation. Current screening methods and gaps are discussed below for both of these phases.

13.3.1 Pre-deployment: Gene Synthesis

Production of a biological weapon with synthetic biology has the potential to be identified and prevented as soon as a nefarious actor tries to purchase DNA fragments from gene synthesis companies. Many U.S. companies conduct background checks on people placing orders and compare the ordered sequence against libraries of listed pathogens as advised by the U.S. National Institute of Health (NIH) (Gronvall 2015; Ahteensuu 2017). If an ordered sequence too closely matches that found within the genome of a listed pathogen then the order isn't fulfilled and authorities may be contacted. However, not all gene synthesis companies follow these steps, and the same recommendations are not in place in other countries (Gronvall 2015). The International Gene Synthesis Consortium (IGSC) is one entity trying to fix this issue by alerting all member companies if one company receives a suspicious order. This prevents malicious actors from simply ordering from another member company if they are denied. However, members of IGSC make up a minority of companies that synthesize DNA but do represent the majority of the market share (Frazar et al. 2017). As a growing number of smaller businesses performing gene synthesis open with increased ease and decreased cost of synthesis, security at this level could become more of a concern. Effort should be made to create similar legislature/recommendations as the NIH in other countries for gene synthesis companies, or to incorporate more of these companies into the IGSC.

Even with the IGSC and national recommendations, however, there is no formal or widely accepted method for prioritizing and listing pathogens (Salerno and Hickok 2007), and the current list of pathogens is not comprehensive (it does not include many agricultural diseases). The list will quickly become insufficient as synthetic biology progresses, and the process of screening gene sequences will become more complicated (Elbers and Knutsson 2013). For instance, scientists have created strains of yeast that can produce opiates and biological toxins (Cirigliano et al. 2017), but it would be illogical to categorize yeast as an agent of concern since it is usually benign and a key research organism. Additionally, malicious actors could hide and later extract desired sequences in a longer sequence or split the sequence among multiple companies to avoid detection (Frazar et al. 2017). Using a list of pathogens to screen for dangerous orders also does not address the encoding of computer malware into DNA as discussed above (Ney et al. 2017). Stronger international policies that require gene synthesis companies to run checks on people placing orders and the orders themselves could be enacted. The government could also monitor the sale and movement of the equipment needed to cut, combine, and propagate gene sequences. The U.S. Department of Commerce currently maintains a list of equipment of concern, but almost none of this equipment relates to synthetic biology (Frazar et al. 2017).

13.3.2 Post-deployment: Pathogen Detection

Once a pathogen is created and deployed, there are ways to identify that an outbreak is occurring and what engineered pathogen is responsible (Trump et al. 2018). Pathogens are often identified using traditional molecular diagnostic methods during natural outbreaks. Some techniques are specific for humans, animals, or plants, but all can be used for an intentional outbreak. Examples include basic and advanced polymerase chain reaction (PCR) technology, fluorescence in situ hybridization (FISH), antibody-based immuno-assays, biochemical testing, mass spectrometry, and enzyme-linked immunosorbent assay (ELISA) (Anand 2018; Krishan et al. 2017). ELISA has been suggested as a viable and rapid way of testing for pathogens/biotoxins released into centralized food processing plants (Wein and Liu 2005). The majority of these techniques, however, must be processed in a laboratory and are constrained by the number of samples they can analyze. Some are also expensive, prone to false-positives, and insensitive to certain bacteria. Optical techniques, such as immunofluorescence (IF), flow cytometry (FCM), thermography, and gas chromatography are useful in detecting diseases in plants (Anand 2018). These too are limited by sensitivity to environmental conditions during measurements and some, like thermography, indicate that a disease is present but cannot diagnose specific diseases.

Synthetic biology can be used to create pathogens that circumvent standard methods of detection, meaning that existing diagnostic tools could return a false negative when analyzing a new synthetically-developed pathogen (Gronvall et al. 2015). For example, the primary way of identifying *Y. pestis* was through the detection of antibodies to the F1 protein found on the outer membrane of the of bacteria. Knowing this, Russia successfully created a strain of *Y. pestis* that lacked the F1 protein in the 1980s, thereby rendering traditional identification methods useless (Zilinskas 2017). This demonstrates that pathogens could be deliberately engineered to avoid available detection techniques, which would delay detection and intervention and subsequently increase the outbreak severity (Gronvall 2015; Plianbangchang 2005). Synthesized pathogens' ability to avoid detection strengthens the argument for ever-evolving detection techniques, in addition to stronger policies around the creation of such pathogens in the first place.

Advancements in the fields of biotechnology, nanotechnology, and information technology should improve detection techniques (Trump et al. 2020b). Nanophytopathology is an emerging technique for the early detection of plant diseases (Anand 2018). Software programs utilizing crowdsourced data and evaluations are also becoming popular for detecting and mapping diseases, and could be used to identify biological attacks against both agriculture and humans (Anand 2018; Berger 2019). The first biological surveillance platform for human outbreaks, ProMed Mail, was created in 1994. It works by sharing and monitoring media, official reports, online information, local observations, and more to identify symptoms or signals that a new or unusual disease outbreak or toxic exposure is occurring (Berger 2019). This program is still alive today and includes over 70,000 members in 185 countries that review reports and engage in discussions on infectious diseases. A number of similar text analysis platforms that compile and analyze databases to identify biological attacks or natural outbreaks have also been launched. These platforms are all affected by the people reviewing the data, the type and language of data sources, region of incidents, and types of cases collected. For instance, the text analysis platform HealthMap identified the 2014 Western African Ebola outbreak a week before WHO declared the outbreak, but a failure in human moderators to recognize the trend identified by the platform caused delay (Berger 2019). Syndromic surveillance can also be used to detect a disease outbreak. Syndromic surveillance uses market data (e.g. the number of eggs being produced, amount of over-the-counter medication being bought, spike in searches about abdominal pain being) to detect when trends differ from the baseline that could indicate a human or agricultural disease outbreak (Elbers and Knutsson 2013; Berger 2019). Overall, the groundwork for early detection systems exists but can be greatly improved upon in the future as advanced computing capabilities increase.

Real-time sensors embedded in the environment that passively monitor for pathogens would be the ideal way of quickly recognizing a biological weapon attack (Shinwari et al. 2014). Biosensors use a recognition element, usually biological in origin, to bind to some component of a pathogen of interest in order to recognize and report that the pathogen is present (Sapsford et al. 2008). These reactions occur in specialized and highly sensitive bioprobes within biosensors. Unlike traditional ways of detecting pathogens, which can take days to complete in a laboratory, bioprobes can recognize specific pathogens within minutes in the field (Kim et al. 2015). One sensor could contain multiple probes focused on specific pathogens and recognize a pathogen from its size, DNA sequence, specific chemical reactions, antibodies, antigens, phage response, or aptamer binding (Petro et al. 2003; Kim et al. 2015). Since pathogens can be modified to avoid specific modes of detection (e.g., the exclusion of certain plasmids or antigens) (Zilinskas 2017), multiple probes using different means of detecting a single pathogen could be included in a single biosensor to increase security (Sapsford et al. 2008). However, screening for specific agents has its limitations, as discussed previously, so biosensors that could recognize novel pathogens or DNA against the natural background would be ideal. Detectors will likely become more accurate and useful as pattern recognition software and nanotechnology advances, as both of these would help to screen out background environmental contaminants, minimize false positives, and decrease detection time (Petro et al. 2003, Kim et al. 2015).

13.4 Conclusion

A biological attack utilizing a synthetic pathogen could prove disruptive to various receptors – making it critical to consider how to prevent, avoid, or ameliorate such challenges *before* they arise (Malloy et al. 2016; Trump et al. 2020d). The threat of attacks using other methods of deployment against less traditional targets (e.g.

technology or the environment) should not be ignored. Increasing synthetic biology capabilities will increase the risk of more complicated biological attacks against all discussed targets, but will also increase our ability to detect pathogens in the environment with biosensors. A large focus, however, should be on preventing the creation of a synthetic pathogen in the first place, which can be accomplished through increased monitoring and regulation of gene synthesis orders, synthetic biology equipment, and distribution of sensitive information.

Literature Cited

Ahteensuu M (2017) Synthetic biology, genome editing, and the risk of bioterrorism. Sci Eng Ethics 23:1541–1561

Anand M (2018) A systems approach to agricultural biosecurity. Health Secur 16(1):58–68

Berger KM (2019) Emerging and enabling technologies. In: Singh SK, Kuhn JH (eds) Defense against biological attacks. Springer, Cham, pp 253–281

Carniel E (2002) The plague. C R Biol 325(8):851–853

Center for Disease Control and Prevention (CDC) (2011) Section 10: Chain of infection in *Principles of Epidemiology in Public Health Practice*, 3rd edn. An Introduction to Applied Epidemiology and Biostatistics. Available at https://www.cdc.gov/csels/dsepd/ss1978/SS1978.pdf

Cirigliano A, Cenciarelli O, Malizia A, Bellecci C, Gaudio P, Lioj M, Rinaldi T (2017) Biological dual-use research and synthetic biology of yeast. Sci Eng Ethics 23:365–374

Cleves PA, Strader ME, Bay LK, Pringle JR, Matz MV (2018) CRISPR/Cas9-mediated genome editing in a reef-building coral. PNAS 115(20):5235–5240

De Vries RP, Peng W, Grant OC, Thompson AJ, Zhu X et al (2017) Three mutations switch H7N9 influenza to human-type receptor specificity. PLoS Pathog 13(6):e1006390. https://doi.org/10.1371/journal.ppat.1006390

Elbers A, Knutsson R (2013) Agroterrorism targeting livestock: a review with a focus on early detection systems. Biosecur Bioterrorism 11:S25–S35

Evans NG, Selgelid MJ (2015) Biosecurity and open-source: the promise and peril of distributed synthetic biological technologies. Sci Eng Ethics 21:1065–1083

Faezi S, Chhetri RR, Malawade AV, Chaput JC, Grover W, Brisk P, Al Faruque MA (2019) Oligo-snoop: a non-invasive side channel attack against DNA synthesis machines. Network and Distributed Systems Security (NDSS) Symposium

Franconi R, Illiano E, Paolini F, Massa S, Venuti A, Demurtas OC (2018) Rapid and low-cost tools derived from plants to face emerging/re-emerging infectious diseases and bioterrorism agents. In: Radosavljevic V et al (eds) Defence against bioterrorism. Springer, Dordrecht, pp 123–139

Frazar SL, Hund GE, Bonheyo GT, Diggans J, Bartholomew RA, Gehrig L, Greaves M (2017) Defining the synthetic biology supply chain. Health Secur 15(4):392–400

Gantz VM, Jasinskiene N, Tatarenkova O, Fazekas A, Macias VM, Bier E, James AA (2015) Highly efficient Cas9-mediated gene drive for population modification of the malaria vector mosquito *Anopheles stephensi*. PNAS 112(49):6736–6743

Getz LJ, Dellaire G (2018) Angels and devils: dilemmas in dual-use biotechnology. Trends Biotechnol 36(12):1202–1205

Gronvall GK (2015) Mitigating the risks of synthetic biology. Council on Foreign Relations, Center for Preventative Action

Gronvall GK, Ravi S, Inglesby T, Cicero A (2015) Singapore-Malaysia-Indonesia-US dialogue on biosecurity. Health Secur 13(6):399–405

Islam MS, Ivanov S, Robson E, Dooley-Cullinane T, Coffey L, Coolin K, Balasubramaniam S (2019) Genetic similarity of biological samples to counter bio-hacking of DNA-sequencing functionality. Nature 9(8684). https://doi.org/10.1038/s41598-019-44995-6

Kelle A, Nixdorff K, Dando M (2010) Strengthening BWC prevention of state-sponsored bioweapons. Bull At Sci 66(1):18–23

Kim J, Gedi V, Lee S-C, Cho J-H, Moon J-Y, Yoon M-Y (2015) Advances in anthrax detection: overview of bioprobes and biosensors. Appl Biochem Biotechnol 176:957–977

Krishan K, Baljinder K, Sharma A (2017) India's preparedness against bioterrorism: biodefence strategies and policy measures. Curr Sci 113(9):1675–1682

Linkov I, Keenan J, Trump BD (2021) COVID-19: systemic risk and resilience. Springer Nature

Malloy T, Trump BD, Linkov I (2016) Risk-based and prevention-based governance for emerging materials. Environ Sci Technol

National Research Council (NRC), Subcommittee on Zinc Cadmium Sulfide (1997) Toxicologic assessment of the Army's zinc cadmium sulfide dispersion tests. The National Academics Press, Washington, DC

Ney P, Koscher K, Orhanick L, Ceze L, Kohno T (2017) Computer security, privacy, and DNA sequencing: compromising computer with synthesized DNA, privacy leaks, and more. USENIX Security Symposium

Noyce RS, Evans DH (2018) Synthetic horsepox viruses and the continuing debate about dual use research. PLoS Pathog 14(10):e1007025

Noyce RS, Lederman S, Evans DH (2018) Construction of an infectious horsepox virus vaccine from chemically synthesized DNA fragments. PLoS One 13(1):e0188453. https://doi.org/10.1371/journal.pone.0188453

Petro JB, Plasse TR, McNulty JA (2003) Biotechnology: impact on biological warfare and biodefense. Biosecur Bioterrorism 1(3):161–168

Plianbangchang S (2005) Strategies of preparedness against the threat of biological warfare and bioterrorism in South-East Asia. Asian Biotechnol Dev Rev 8(1):77–98

Salerno RM, Hickok LT (2007) Strengthening bioterrorism prevention: global biological materials management. Biosecur Bioterrorism 5(2):107–116

Sapsford KE, Bradburne C, Delehanty JB, Medintz IL (2008) Sensors for detecting biological agents. Mater Today 11(3):38–49

Shinwari ZK, Khalil AT, Nasim A (2014) Natural or deliberate outbreak in Pakistan: how to prevent or detect and trace its origin: biosecurity, surveillance, forensics. Arch Immunol Ther Exp 62:263–275

Shipman SL, Nivala J, Macklis JF, Chruch GM (2017) CRISPR-Cas encoding of a digital movie into the genomes of a population of living bacteria. Nature 547:345–349. https://doi.org/10.1038/nature23017

Sobel J, Khan AS, Swerdlow DL (2002) Threat of a biological terrorist attack on the US food supply: the CDC perspective. Public Health 359:874–880

Trump BD, Foran C, Rycroft T, Wood MD, Bandolin N, Cains M et al (2018) Development of community of practice to support quantitative risk assessment for synthetic biology products: contaminant bioremediation and invasive carp control as cases. Environ Syst Decisions 38(4):517–527

Trump BD, Cummings CL, Kuzma J, Linkov I (2020a) Synthetic biology 2020: Frontiers in risk analysis and governance. Springer

Trump BD, Galaitsi SE, Appleton E, Bleijs DA, Florin MV, Gollihar JD et al (2020b) Building biosecurity for synthetic biology. Mol Syst Biol 16(7):e9723

Trump BD, Keisler JM, Galaitsi SE, Palma-Oliveira JM, Linkov I (2020c) Safety-by-design as a governance problem. Nano Today 35:100989

Trump BD, Keisler JM, Volk KM, Linkov I (2020d) Biosecurity demands resilience. Environ Sci Technol

Wein LM, Liu Y (2005) Analyzing a bioterror attack on the food supply: the case of botulinum toxin in milk. PNAS 102(28):9984–9989

Yeh J-Y, Park J-Y, Cho YS, Cho I-S (2012) Animal biowarfare research: historical perspective and potential future attacks. Zoonoses Public Health 59:536–544

Yeh J-Y, Park J-Y, Cho YS, Cho I-S (2013) Countering the livestock-targeted bioterrorism threat and responding with an animal health safeguarding system. Transbound Emerg Dis 60:289–297

Yong E (2018) A controversial virus study reveals a critical flaw in how science is done. The Atlantic. Available at https://www.theatlantic.com/science/archive/2018/10/horsepox-smallpox-virus-science-ethics-debate/572200/

Zilinskas RA (2017) A brief history of biological weapons programmes and the use of animal pathogens as biological warfare agents. Scientific and Technical Review of the Office International des Epizooties 36(2):415–422

14

Biosecurity Governance and the Role of Tripwires

Benjamin D. Trump, Stephanie Galaitsi, Miriam Pollock, Kaitlin M. Volk and Igor Linkov

14.1 Introduction

Synthetic biology is a new, rapidly growing interdisciplinary field which seeks to use engineering techniques to alter and construct new biological components, devices, and systems. The rate of synthetic biology development and research has increased over the past two decades in both industry and academia (Ahteensuu 2017). Applications include medicine (new vaccines, delivery of therapeutics, and treatments), energy (biofuels), environmental remediation, food production, and general industry (detergents, adhesives, perfumes) (Evans and Selgelid 2015; Gronvall 2015). While synthetic biology heralds advances in these fields, its techniques could also be adapted for malicious purposes and used by terrorist organizations, rogue actors, or hostile nations to create dangerous pathogens, invasive organisms, or other disruptive biological agents (Yeh et al. 2012). Such potential makes synthetic biology a dual-use research area of concern (DURC) as the same techniques can be used to benefit or harm people, animals, environments, technology, or nations (Getz and Dellaire 2018). To cope with threats arising from synthetic biology's dual-use nature, biosecurity is needed to prevent, detect, and attribute biological attacks.

The rapid development of synthetic biology is lowering barriers to access its technological capabilities. While organizations seek to develop and formalize effective governance strategies for managing the technology's risks to humans and the environment, an increasing number of independent actors are engaging in the technology (Trump et al. 2020). Existing biosecurity policy practices have left gaps in biosecurity effectiveness for synthetic biology. Such gaps require an iterative

---

B. D. Trump (✉) · S. Galaitsi · M. Pollock · K. M. Volk · I. Linkov
US Army Corps of Engineers, Concord, MA, USA
e-mail: Benjamin.D.Trump@usace.army.mil

response to influence countries and organizations away from harmful or nefarious synthetic biology research applications.

In biosecurity, triggering mechanisms that signal when a nation, sub-state actor, or individual is on the path towards biological weapons development and deployment can be part of this response. These mechanisms, called tripwires, are actions, events, or breakthroughs that impel a country toward either enabling or moving away from biosecurity threats. They can include both bottom-up and top-down signals. Bottom-up tripwires include information hazard (the release of information by an individual, company, or academic institution that could be used by a malicious actor), profit motivation, and partnership with entities of biosecurity concern (whether low or high). Top-down tripwires also include information hazard, as well as internal or external conflict that could motivate a nation to develop and use biological weapons to maintain power, a public health crisis that could cause biosecurity and biosafety regulations to be relaxed in search of a treatment or cure, and geopolitical alignment with a nation of some degree of biosecurity concern.

14.2 Biological Weapons in History

Biological weapons have been used since ancient times. These ancient weapons included scorpion bombs (terracotta pots filled with deadly scorpions and thrown at the enemy), crushed toxic hellebore plants, and arrows tipped with poison (Mayor 2003). As states sought new ways to generate advantages over their opponents, they developed increasingly advanced biological weapons. In 1925, in large part as a response to the devastating use of chemical and biological weapons in World War I, the Geneva Protocol was passed, banning the usage of biological and chemical weapons in wartime. Nevertheless, many states continued their biological weapons programs. During World War II, for instance, Japanese units deployed bombs filled with plague-ridden fleas in China as a part of their biological weapons program.

In 2001, in the worst biological attack in US history, letters laced with *Bacillus anthracis* spores, or anthrax, were sent through the U.S. Postal Service (DOJ 2010). The so-called “Amerithrax” bioterror attack resulted in five deaths, 22 infections, the prophylaxis of over 30,000 people, and hundreds of millions of dollars in decontamination expenses, as well as around $6 billion in lost revenue and new security measures (DOJ 2010). After a lengthy investigation, it was determined that one man, acting alone, had been responsible for the attack. This illustrates the enormous destructive potential of bioweapons (FBI n.d.; Ellis 2014).

Many countries, including the US, UK, and Russia, were developing biological and chemical agents for warfare until the Convention on the Prohibition of the Development, Production and Stockpiling of Bacteriological (Biological) and Toxin Weapons and on Their Destruction was signed in 1972 (“BWC”). Since 1970, there have been 35 confirmed cases of biological weapon deployment (Franconi et al. 2018), with the majority of research and weapons production being pursued by state actors. However, synthetic biological weapons are expected to become a larger concern as the field advances and as malicious sub-state or non-state actors pursue and adopt these weapons (Gronvall 2018).

For almost 50 years, the BWC has been the gold standard for watching for and preventing state-sponsored biological weapons. However, advancements in synthetic biology have decreased the time, money, and skill needed for an individual or organization to participate in the field, leading to the development of pathogens by smaller sub-state and non-state actors who are not restricted by the BWC (Evans and Selgelid 2015). Some interested individuals and small companies interested in synthetic biology have set up labs in their homes using household items and equipment or kits ordered over the internet in a movement now known as "DIY Bio" or "garage biology" (Ledford 2010). Additionally, while chemicals have a set structure and formula that makes them easier to identify and govern, advancements in synthetic biology allow for the creation of novel types of biological weapons that are not explicitly prohibited by the BWC but instead exist in a sort of "gray zone." This increased accessibility of synthetic biology and difficulty in categorizing synthetic biology products creates a threat space that is not fully captured within the structure or implementation of the BWC.

Synthetic biology raises the possibility that pathogenic bioweapons could be designed, developed, and deployed in new ways that diverge from the disease-causing characteristics of naturally occurring pathogens. Traditionally, only known pathogens found naturally in the environment, such as *B. anthracis* and *Y. pestis*, were developed as biological weapons because of the inherent infectious characteristics that readily enabled such organisms to serve as weapons. However, as synthetic biology continues to expand capabilities to create and modify biological weapons, there is an increasing need for biosafety and biosecurity assurances for humans, animals, plants, and the environment. The following sections examine the current and future state of research on five categories of agents, or "platforms," that could be used as biological weapons: viruses, microbes, multicellular organisms, cell-free systems, and prions.

14.3 Promoting Responsible Research in Modern Biotechnology: Tripwires as a Framework to Understand Red-Team Capacity and Intent for Nefarious Deployment of Technology

States will pass through a series of broad stages on the path to developing and deploying biological weapons. There are characteristics typical of states and actors at each stage, and these characteristics can serve as tripwires to identify if a group is poised to become more or less of a biosecurity threat. We identify four stages: (1) the indication of an interest in synthetic biology, (2) the achievement of scientific, technological, and engineering capacity for synthetic biology, (3) the development or acquisition of synthetic biology weapons, and finally, (4) the deployment of synthetic biology weapons. Stage 1 is further broken down into top-down and bottom-up initiatives, while Stage 3 is broken down into homegrown development of weapons and acquisition from external sources.

This section concludes with a discussion on technological tripwires, or advancements in the techniques and technology used in synthetic biology which will drastically increase what people are capable of doing with synthetic biology. A summary of different tripwires can be found in Table 14.1.

14.3.1 Regime Characteristics that Could Lead to Synthetic Biology Weapons Development and Their Tripwires

This section provides a preliminary process model, developed through the distillation of various country profiles, of the steps leading to the deployment of synthetic biology weapons. This process model lists the outward signs a country can project in its path to weapons development, and proposes actions that might be taken to

Table 14.1 Categories and descriptions of tripwires that can be used to indicate when a nation, sub-state group, or individual is becoming more or less of a biosecurity threat

Top-down	
Information Hazard	The release of information by a nation that could be used by a malicious actor, or the suppression of bottom-up information hazards
Conflict	Internal or external conflict that motivates a nation to develop and use biological weapons to maintain power
Public health crisis	A natural epidemic that causes biosecurity and biosafety regulations to be relaxed in search of a treatment or cure
Geopolitical alignment	Establishing relations with a nation that has a record of biosecurity compliance or incompliance
Bottom-up	
Information Hazard	The release of information by an individual, company, or academic institution that could be used by a malicious actor
Profit motivation	The perception that money can be made through the use of synthetic biology
Partnership	Joining with a high or low biosecurity threat entity in new partnerships, mergers, or research consortia
Technological	
Oligonucleotide synthesis	Synthesis of nucleotides into the desired sequence
DNA assembly	Assembling multiple smaller fragments of oligonucleotides into the desired larger sequence
Genetic modifications	Editing and inserting desired sequences into targeted sites of a genome
Protein engineering	Designing and producing proteins and other biomolecules in a laboratory with natural or unnatural parts

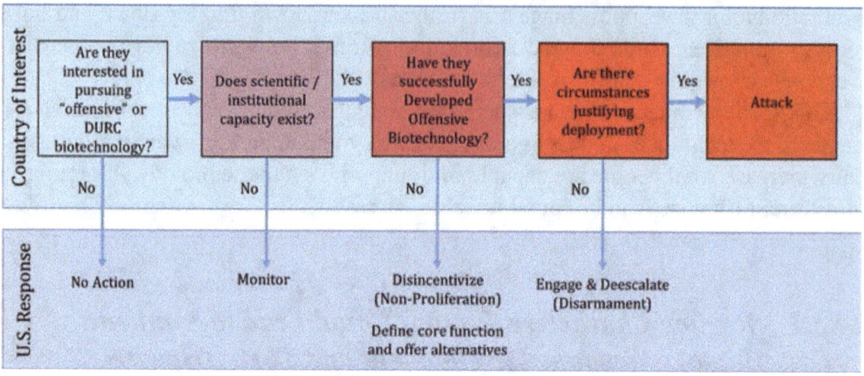

Fig. 14.1 Demonstrative synthetic biology biosecurity event tree

engage with countries at each stage of the process. Figure 14.1 shows an example process model, in the form of an event tree, which might be used by the US.

This section also lists potential tripwires that can be used to identify when a country is moving into a new stage and becoming more or less of a biosecurity threat. These can include alterations in a nation's (a) incentives to engage in Dual-Use Research of Concern (DURC), (b) degree of security and control over institutional and individual activity in the synthetic biology space, and/or (c) potential for a militarization of synthetic biology and its enabling technologies.

14.3.1.1 Stage 1a. Bottom-Up Initiative: Independent Actors Indicate Interest in Synthetic Biology

Corporate, academic, and other non-governmental actors may have incentives to pursue the development of synthetic biology. In this stage of the process model, the economic incentives can indicate the value of various non-weapon applications of synthetic biology, including in the fields of energy, pharmaceuticals, agriculture, and medical applications. Thus, any monitoring of the synthetic biology domain during this stage will require an understanding of any influential economic fields that could benefit from synthetic biology development, particularly fields with high Dual-Use Research of Concern (DURC) applicability. Profit motivation is a large tripwire for signaling a bottom-up interest in synthetic biology. If individuals or institutions perceive that a profit can be made through the use of synthetic biology, then they are more likely to become interested.

Bottom-up actors interested in developing synthetic biology may also be responding to an impetus other than economics. For example, academics who studied abroad may wish to continue their research by opening a synthetic biology laboratory in their home country. Without the direct input of the government, bottom-up development is more likely to be directly related to the ambitions and profits of small academic or commercial entities, both of which have low incentives to prioritize weapons development and high incentives to be transparent about their work (especially during the beginning stage when they might be acquiring important partners or support).

Information hazards are a tripwire to watch for groups moving from an interest in synthetic biology to achievement of synthetic biology capabilities (Stage 2). An information hazard can be defined as: "A risk that arises from the dissemination of (true) information that may cause harm or enable some agent to cause harm." Such a hazard may arise through data spillage of classified and/or sensitive information related to scientific and/or technological capabilities to acquire, engineer, test, and/or build DURC-related research and products. Typically, an information hazard is seen as the "critical cornerstone on the bridge from inspiration to actualization" of a biological weapon or harmful engineered biological product (Esvelt 2018; Kirkpatrick et al. 2018).

In order to address this, some countries have a strong culture of "self-governance," whereby teams of scientists proactively tackle DURC concerns before, during, and after the completion of a research project. However, other states (in particular, many East Asian and Southeast Asian nations) lack this custom, instead expecting biosecurity management to come from powerful government officials who have limited knowledge of or expertise in the research being performed. The publication of DURC with little legitimate commercial and/or social utility (i.e., the recombination of extinct human pathogens) should be monitored, and actions to reduce the occurrence of such publication should be taken at the institutional, editorial, or state level. New partnerships or mergers between companies and/or universities, or the creation of research consortia, the membership of which comprises representatives from multiple nations, could indicate the risk of a bottom-up information hazard leading to a nation gaining synthetic biology capabilities. Monitoring should be conducted to see if the new partnerships or mergers include members who would increase biosecurity (e.g., watchdog NGOs and/or participants with an established record of compliance with the BWC) or decrease it (e.g., participants with a record of noncompliance).

Countries currently developing bottom-up initiatives in synthetic biology include Iran, where academics have begun attending international topical conferences, Saudi Arabia, where the King Abdullah University has a program in synthetic biology, and Ukraine, where several universities offer coursework in synthetic biology.

Bottom-up initiatives in synthetic biology do not indicate a lack of support from the governing regime, only a lack of coordinated efforts. Bottom-up initiatives can also provide impetus for top-down initiatives (Step 1b).

14.3.1.2 Stage 1b. Top-Down Initiative: Regime Indicates Interest in Synthetic Biology

A coordinated top-down effort to support synthetic biology research can emerge from a governing regime independent of whether bottom-up activities are already occurring. Because synthetic biology is a DURC technology, any regime publicly signaling its interest in developing synthetic biology capabilities has potentially taken an early step on the path of synthetic biology weapons development. The funding and other incentives made available could hypothetically be used to direct research in a way that benefits the government itself, including through weapons development. However, many products of synthetic biology are not and cannot be used as weapons: interest in synthetic biology does not necessarily indicate nefarious intentions or the desire to develop a weapons program. Accordingly, the economic context of research ambitions and their relationship with other regime ambitions is highly relevant to any evaluation of the government's intentions. Monitoring of countries in this stage should relate the synthetic biology ambitions of the regime to other stated ambitions, such as exports they may seek to expand. In this way, profit motivation is also a top-down tripwire for indicating when a regime is likely to become interested in pursuing synthetic biology.

Information hazards are also an important tripwire for a top-down initiative moving from interest to capacity. For synthetic biology, Esvelt and Kirkpatrick et al. note the importance of states as protectors against information hazards that could inspire rogue militaries, terrorist groups, companies, or even individuals to learn and pursue synthetic biology research. Increasingly, however, the capacity for centralized governments to regulate biosecurity information hazards is becoming prohibitively difficult due to the globalized and increasingly diversified nature of synthetic biology research (Trump et al. 2020).

14.3.1.3 Stage 2. Achievement of Scientific, Technological, and Engineering Capacity for Synthetic Biology

Following an expression of interest and subsequent investment in synthetic biology, groups or individuals within a country may successfully develop synthetic biology applications. During this stage of synthetic biology development, monitoring should occur. Academic publications as well as commercial patents, products, and processes are worthy of being monitored, as they can all facilitate greater understanding of a country's internal synthetic biology capabilities. Exchanges of human capital are another source of information, though this may be more difficult to observe or monitor, especially within corporate structures. It is, however, possible that although corporate profit motivations cause actors to conceal specific details of their work, the overall capabilities are still evident in the final products.

Tripwires that would indicate a country at this stage as being more or less of a biosecurity threat include public health crises and geopolitical alignment. Public health crises can take the form of epidemics, sustainability crises, pollution, or other events where public health is directly threatened by a catalyst. Whether targeting humans, animals, crops, or the natural environment, epidemics represent the most visible and urgent of public health crises relevant to biosecurity. In such crises, states are incentivized to bend or relax established rules and norms regarding biological safety in search of vaccines, treatments, and cures for the disease. Such a relaxation of biological safety standards could become a biosecurity concern. The pressure and financial incentives to foster interventions for public health crises can drive institutional actors to ignore biosecurity rules, share sensitive information that may be applied to DURC, or otherwise facilitate the development of DURC products and materials that would otherwise, in non-emergency situations, be more closely scrutinized.

Geopolitical alignment refers to the alliances, agreements, or participations to which a state voluntarily commits itself. The development of shared scientific research or collaborative agreements between states can signal movement in the strength of a state's biosecurity intentions and enforcement capabilities, both directly and indirectly. Tripwires relating to geopolitical alignment include direct participation within collaborative security agreements like the BWC, which would indicate a state's biosecurity threat is reduced, or indirect partnerships with other states that have a strong or weak biosecurity record. The potential increased capabilities and reduced biosecurity-biosafety standards indicated by this tripwire represent a hazard that can benefit non-state actors as well.

During this stage, monitoring should encompass laboratory construction, whether private, academic, or governmental, with special attention to features in construction that suggest clandestine laboratory spaces, as well as any synthetic biology funding sources and their disbursements. The exchange of human capital between countries with different competence levels can help indicate the direction of research and future capabilities, and the quality of both bottom-up and top-down biosecurity initiatives will reveal the ease with which researchers could skirt biosecurity guidelines, regulations, and laws for their own benefit. Finally, the published outputs and marketed products will provide insight into the level of synthetic biology capabilities within the country. Economic incentives, including potential clients for weapons exports, should continue to be monitored.

14.3.1.4 Stage 3a. Active Development of Synthetic Biology Weapons

Once competence in synthetic biology is achieved, it is possible that some synthetic biology research may be directed into national laboratories for state-funded weapons development purposes. These laboratories do not have the same transparency practices as academic or corporate actors, and it should be assumed that such weapons development will not be reported to the international community. Monitoring of countries in this stage may include tracking the flow of experienced human capital from for-profit or academic sectors into government sectors, with special attention to gaps or abrupt decreases in publications. This could indicate clandestine synthetic biology development under regime funding, which could indicate offensive purposes.

Additionally, independent actors may pursue weapons development outside of their government institutions, if provided the correct incentives by patrons. Given access to sufficiently powerful technology, independent actors may harness it for nefarious purposes if the benefits would be useful or worthwhile to them and exceed the benefits of abiding by existing biosecurity frameworks. Similarly, profit motivation could also lead countries and sub-state actors to develop biological weapons at the behest of a partner nation or client, making profit motivation a tripwire for entering this stage.

A state's record of internal strife and recent or ongoing external conflict can be used as another tripwire for a state or sub-state actor's willingness to develop a synthetic biology weapon or to move to Stage 4, the deployment of such a weapon. While a past record of such conflict can help characterize a state's likelihood to embrace biological weapons, its position in ongoing or potential conflict in the near term serves as an especially significant tripwire of concern. External conflict, or active combat between two or more states, is a tripwire of concern as it incentivizes a government to pursue force multipliers or scientific innovations that will provide an edge in combat. Such a tripwire is particularly enticing for middle income states that are currently fighting or may fight a state of comparable or even superior power. For example, Iraq under Saddam Hussein deployed chemical weapons against poorly-armed yet highly-motivated and organized Iranian forces (Ali 2001; Szinicz 2005). Iraq increased its use of unconventional weapons as the war dragged on and as conventional Iraqi armor, aircraft, machines, and weaponry were exhausted.

For external conflict, biological weapons represent (a) a first-strike capability, capable of crippling enemy health, armor, or agriculture, (b) a force multiplier, capable of providing an edge against a near evenly-matched foe, or (c) a survival mechanism to attack an enemy who has the upper hand in a conventional war, where the losing party becomes increasingly desperate to fend off attacks and ensure survival. Internal conflict includes, among other scenarios, civil war, riots, and other forms of civil strife or internal dissension. Chemical and biological weapons have been considered and/or deployed as a means of (a) instilling fear and control over parties in dissent against the regime, or (b) as a force multiplier to inflict mass casualties against an overwhelming rebellious force. As an example of (a), the government of Syria – which began to seek unconventional weapons in the 1990s – deployed chemical weapons as Syrian Government forces began to lose ground in the Syrian Civil War (Eisenkraft and Falk 2019; Diab 1997). An example of (b) is the apartheid government in South Africa's active development, testing, and deployment of biological weapons to assassinate key anti-government figures and to serve as a weapon of last resort to quell a mass revolt against the government (Hay 2016).

Monitoring efforts for countries in this stage may include monitoring imports, exports, information transfers, and general geopolitical relationships with other governments, especially those engaged in belligerence elsewhere, either domestically or internationally. Efforts should also include monitoring whether a state or non-state actor has or will be challenged by a powerful force against which conventional armaments may be insufficient to ensure survival, or if there is the potential for widespread civil strife or civil war. Sufficient indications of clandestine synthetic biology development at this stage may necessitate active interventions rather than passive monitoring, though the monitoring specified in Stage 2 should also continue. Interventions could include overtures to encourage the state to join international biosecurity efforts like the BWC and UN Resolution 1540 if it has not done so already, or to submit to routine or supplemental monitoring according to the circumstances. There are currently no countries that have publicly admitted to pursuing or possessing synthetic biology weapons capabilities.

14.3.1.5 Stage 3b. Acquisition of Synthetic Biology Weapons from an External Source

Countries or non-state actors without synthetic biology capabilities may choose to forgo developing domestic technological capabilities if they are able to acquire synthetic biology weapons through other avenues. This requires access to competence elsewhere, either through existing relationships or through building new ones. Regimes possessing advanced technological capabilities may attract actors with inclinations to learn and then use these capabilities for violence, as was the case with the 9/11 hijackers who trained in U.S. flight schools.

With DURC technologies, weapons development can arise inadvertently through exploration of beneficial uses for innovations. In contrast, actors willing to invest only in weaponized outputs signal a lack of economic interest in peaceable applications for the technology, and for this reason their interest in the weapons applications may indicate a stronger propensity to ultimately deploy them. However, such weapons could also serve as a tool of deterrence against another regime, though this would necessitate publicizing capabilities at some point.

It may be difficult to determine when an actor who is aiming to purchase synthetic biology weapons ultimately acquires them. However, regardless of how far along in the process of acquiring these weapons an actor is, situation monitoring can entail disincentives for armament, whether economic assistance or coercive or offensive actions. Such actors may be best identified through their relationships with competent synthetic biology developers, whose outputs, collaborators, and clients should already be being monitored according to Stage 2. If a particular vendor appears to be on the cusp of selling a weapon, engagement can involve encouraging the vendor to increase their biosecurity processes as well as public engagement with the deal, since the public may be uncomfortable with selling synthetic biology weapons, especially to the type of belligerent actors likely to seek them.

14.3.1.6 Stage 4. Deployment of Synthetic Biology Weapons

Whether weapons are produced domestically or acquired from external sources, an instigating event will likely need to occur before their deployment. History abounds with examples of events that have triggered offensive actions, including both acute events (the secession of the Confederacy, which started the US Civil War) and chronic events that built over time (as resistance to the Assad regime of Syria mounted, the regime deployed chemical weapons). The nuclear bombs that ended World War II in the Pacific were part of a larger Allied strategy, but the timing of the deployment was more determined by weapons development (Stage 3a) within an extended conflict than a single precipitating factor. Still, even if there is not an acute precipitating event, a regime considering deploying weapons must both have capability and a belligerent relationship with another entity. Thus, any regimes already under surveillance or being engaged according to Steps 2 and 3 should be further engaged in the event of a conflict erupting.

We propose a specific profile for actors at risk of deploying synthetic biology weapons, which can constitute a means to identify them and evaluate their risk levels. In addition to their geopolitical alignment, which determines whether they have access to the weapons and a reason to deploy them, such actors are likely to be fairly risk tolerant because the spatial and temporal effects of a synthetic biology attack are not currently well-understood. Non-scientist citizens are often averse to new, potentially harmful technologies, and regimes that give those citizens more voice in governance risk paying a high price for such weapon deployment. Citizens from regions that have already suffered the ill effects of scientific advancements may be especially averse to deployment. For example, Ukrainians and Belarussians, who were especially negatively affected by the Chernobyl disaster, and Kazakhis, who have suffered from above-ground nuclear testing sites, may be particularly averse when it comes to the deployment of shadowy, poorly-understood novel weapons. Similarly, countries exhibiting cautionary principles towards genetically modified organisms will be less likely to resort to synthetic biology weapons because of pre-established risk aversion.

However, the opinions of the population only matter if the regime in possession of the weapons is answerable to that population. A risk-tolerant regime may be one characterized more by authoritarianism than democracy, or it may be a group of non-state actors without demarcated constituents. Therefore, one metric to estimate the probability of synthetic weapons deployment could arise from examining a regime's political participation and human rights record, which indicates its ability to ignore or dehumanize segments of the population. Another metric may be whether the regime's rhetoric somehow presents itself as insular and separate from the targets of a potential weapons deployment. Reality is frequently subservient to perception, and in moments of high stress, actors who do not answer to a larger and more moderating population and are accustomed to dehumanizing their adversaries may deem the benefits of synthetic biological weapons to be worth the uncertain risks. Therefore, there is a spectrum of likelihood in weapons deployment that depends on various factors beyond weapons possession. Finally, decision makers will weigh these factors against the magnitude of the damage they anticipate if the weapon is not deployed. If they are facing a real or perceived existential threat, they may see no reason for restraint.

Should events proceed to the point where a regime considers launching a synthetic biological weapon, active engagement will be urgently needed. This could include attempts to physically compromise the weapon and the deployment infrastructure, or diplomatic efforts that assure the regime of a sufficiently positive outcome to its conflict without launching the weapons. The latter strategy resembles that used by the US during the 1973 Arab-Israeli War, in which the US quickly intervened on Israel's side in order to sway the tide after Israel threatened to use nuclear weapons. Ideally, such a last-ditch effort would only be necessary after the failure of many other efforts to diminish synthetic biology weapons capacity starting from the time when the actor in question entered hostilities with another entity. This also raises the possibility that governments might threaten to deploy synthetic biology weapons in order to receive assurances. In these cases the information gleaned from earlier monitoring should be able to inform the actual veracity of the threat. Additionally, if sufficient actions are taken in earlier stages of weapons development, such a situation should not arise.

14.3.2 Key Enabling Technologies and Technological Tripwires

Many enabling technologies are key to further progress in synthetic biology. Just as synthetic biology enables potential misuse with dual-use studies or technologies, developments in the technologies or techniques outlined below also allow for dramatic growth of potential security threats by expanding the capabilities of synthetic biology, and therefore the abilities of actors. Understanding the key breakthroughs needed for the expansion of synthetic biology capabilities is necessary for regulatory agencies to anticipate and respond to potential and imminent threats. If regulatory agencies are aware of the relevant technologies and where key breakthroughs are likely to occur, signs of advancements in these technologies can act as tripwires and signal to defense agencies when new threats have emerged due to the now-enhanced capabilities of synthetic biology and malicious actors. It is important to note that as with nearly all aspects of synthetic biology, enhanced synthetic biology capabilities are not in and of themselves normatively "bad"; they will overwhelmingly be put to beneficial uses. However, such technological advancements will also open the door to the creation of more complex or novel biological weapons that aren't possible with today's technology, and thus these advancements should be monitored so that novel types of biological weapons can be continually anticipated and defended against as they are developed.

It is also critical to acknowledge the increasing usage and development of computing power and Artificial Intelligence (AI) in scientific advances. Synthetic biology harnesses these resources to improve many different aspects of product production, including fidelity and automation. With major advances in computing and AI, the capabilities of synthetic biology will also expand. While not explicitly mentioned above as a technological tripwire, computing power serves as an undercurrent of development that cannot be ignored. Advances in these particular fields will produce the most radical changes in synthetic biology capabilities and can thus most easily serve as tripwires for observing the capacities of synthetic biology as a whole and subsequently the capabilities of any "malicious actors."

14.4 Conclusion

Building an effective biosecurity strategy to encompass twenty-first century synthetic biology requires understanding the novelties synthetic biology creates in the biosecurity threat space, as well as the structural vulnerabilities it can exploit and the likely causes of inadequate biosecurity practices. Tripwires can be helpful in alerting monitoring groups and governments as to potential biosecurity threats. Synthetic biology's novel biosecurity concerns arise from its broad scope, wider availability, complexity, and uncertainty over current and future capabilities. Critical developments such as gene editing via CRISPR vastly improve upon previous genetic engineering processes and may yield a revolution in human and environmental health research, but may also cause substantial and irreversible harms.

Harms may also arise through the deliberate use of such techniques to disrupt human and environmental systems. Misuse of synthetic biology techniques requires two circumstances: (a) the spread of ideas, techniques, or knowhow to utilize synthetic biology's enabling technologies for irresponsible or nefarious purposes ("information hazard"), and (b) the ability to use such knowledge and tools to generate and disseminate harmful engineered organisms to vulnerable recipients. The tripwires we have outlined in this chapter aim to signal when these necessary circumstances have been or are close to being reached.

Synthetic biology advancements have already included the publication and dissemination of a methodology for synthesizing horsepox in a laboratory setting. Some critics say this information could support a nefarious actor to reconstitute and develop smallpox, or to synthetize other viruses. Additionally, the widely publicized recreation of the 1918 Spanish Influenza, which killed some 50 million people worldwide at the close of World War I, could facilitate the synthesis process for actors wishing to cause harm (Tumpey et al. 2005). Even nonpathogenic approaches have been defined as dual-use research, ranging from the disruption of local ecologies via gene drives to the manipulation or destruction of inorganic materials.

These and dozens of other cases demonstrate the increasing ease by which an actor can acquire information and apply existing tools to deploy advanced genetic engineering applications with limited to no oversight. In 1975, the U.S. National Institute of Health (NIH) established compliance measures for genome engineering that were enforced through funding restrictions; however many synthetic biology innovators in the US can now operate without NIH funding, approval, or even awareness, and the NIH does not oversee research in other countries. Today, the financial costs, time limitations, and skill requirements needed to wield synthetic biology tools have scaled down to become accessible even to high school students. Furthermore, the required baseline knowledge will further diminish over time as synthetic biology processes become more streamlined. While such broad access to sophisticated genetic engineering knowledge and equipment can accelerate scientific breakthroughs, it also places the responsibility of biosecurity on a huge number of unsupervised actors across the globe.

In 2018, the states-parties to the BWC noted that increased access to technologies such as gene editing, gene drives, and gene synthesis is being conferred to actors with limited or no oversight from established industry or governmental organizations, raising concerns about potential violations of the BWC. It is helpful to forecast and understand looming threats and potential mitigation strategies at various scales, but international treaties are not structured to oversee bottom-up efforts related to the localization and globalization of synthetic biology below the national scale. One response may be the broadening of engagement from traditional oversight agencies like NIH. Where top-down governance proves insufficient, other actors such as universities, non-profits, and companies will need to engage their gatekeeping and watchdog capabilities to protect against nefarious actors. Top-down governance may support such initiatives, which will require harmonization and communication up to the international level. Tripwires, in the form of a game theoretic approach, can be critical in identifying the range of timelines whereby nefarious deployment of biotechnology becomes realistic and actionable.

Using the lens of risk analysis, longstanding biosecurity policy practices appear to have produced gaps in biosecurity effectiveness for synthetic biology. These policies include (a) the framing of security as a cost or undesirable expense to be minimized, (b) the siloing of scholarship and practice across disciplinary domains and between academia, government, industry, and civil society, and (c) the narrow framing of security issues that ignore technological developments occurring within a number of different countries, by new actors, and/or in adjacent technology fields. Each of these concerns could be addressed by policy solutions that both support technological development and mitigate security threats while facilitating public engagement in synthetic biology and investment in its products as they enter the marketplace. Policies for synthetic biology must be scalable, transferable, and adaptable in order to address its emerging technical and social challenges.

Advanced biological research is no longer overwhelmingly dominated by Europe and the US, and this may introduce different approaches to or priorities for biosecurity. For instance, Russia's Federal Research Programme for Genetic Technologies Development for 2019–2027 intends to "implement a comprehensive solution to the task of the accelerated development of genetic technologies, including genetic editing; to establish scientific and technological groundwork for medicine, agriculture and industry; to improve the system of preventing biological emergencies and monitoring in this area." Similarly, Saudi Arabia is funding research related to the development of microbial cell factories to produce fuels and chemicals. Meanwhile, the Singaporean government is investing considerable resources into the funding of life and environmental sciences research at Nanyang Technological University, the National University of Singapore, and the Agency for Science, Technology and Research (A*STAR). The Chinese Academy of Sciences is establishing an Institute of Synthetic Biology, which is tasked with the dual responsibilities of fostering roadmaps for the future development of Chinese synthetic biology while also establishing safety and security norms for researchers at Chinese institutions. There are no top-down efforts beyond existing mechanisms like the BWC or the CWC to standardize global governance and usage of synthetic biology, and bottom-up efforts are not coordinated in their reach or messaging.

Relative newcomers to synthetic biology development may possess differing tolerances and constructions of risk than more established technology developers. The implications, though vast, can be grouped into two general areas. One includes diverging safety and security practices at various points of an international supply chain that forms the backbone of an increasingly globalized economy. Another includes the potential for small-scale experiments or national biosecurity policies to escape a given actor's control and spill across political boundaries. While one country may find the environmental risk of a particular synthetic biology application acceptable, its spread across borders into another country may disrupt those local ecologies (i.e., crashing or hardening a particular species through genetic engineering) or expose vulnerable human populations to irreversible consequences without options for amelioration. The nature of certain synthetic biology applications (i.e. gene drives) makes it impossible for risk averse countries to wholly quarantine themselves from exposure to harms resulting from another country's decisions. This is also an issue of equity because risk-tolerant countries will reap the rewards when beneficial technologies emerge, but risk-averse countries may be forced to bear their neighbors' risks without any means to capture potential rewards.

An environment of competing and incongruent risk architectures causes individual states, organizations, or industries to arrive at differing definitions of security threats or acceptable levels of loss in pursuit of a technology's intended gains. For a technology as uncertain as synthetic biology, this policy divergence may set governments, companies, and other research organizations down vastly differing policy paths, and impede consensus in assessing the minutiae of technical risk concerns or assessment protocols, or ensuring security for anyone.

Effective global biosecurity will not happen quickly, nor will it be enthusiastically adopted by all national governments or non-governmental organizations. Incentives to misuse synthetic biology with harmful consequences remain high for certain negligent actors, as are the incentives to dual-use by nefarious actors, and the coming years may see such misuse affecting human or environmental health. Successful biosecurity implementation must be adaptable to quickly incorporate uncertainty as well as new capabilities. Urgent steps are required to place such notions into practice before a major threat incident, both to prevent the damage incurred by such an incident and to avoid subsequent policy changes which could limit or ban platforms of synthetic biology research entirely. Besides forgoing the benefits of the technology, such a ban could force development underground and further out of the reach of coordinated governance or risk assessment. By identifying the social, economic, institutional, and technological tripwires that influence a state's trajectory towards biosecurity research of concern, responsible parties can take steps to apply biosecurity to maximize technological benefits while minimizing its dual-use potential.

References

Ahteensuu M (2017) Synthetic biology, genome editing, and the risk of bioterrorism. Sci Eng Ethics 23(6):1541–1561

Ali J (2001) Chemical weapons and the Iran-Iraq war: a case study in noncompliance. Nonproliferation Rev 8(1):43–58

Biological Weapons Convention (1972) United Nations

Department of Justice (2010) Justice Department and FBI announce formal conclusion of investigation into 2001 Anthrax attacks. Press release, February 19. https://www.justice.gov/opa/pr/justice-department-and-fbi-announce-formal-conclusion-investigation-2001-anthrax-attacks

Diab MZ (1997) Syria's chemical and biological weapons: assessing capabilities and motivations. Nonproliferation Rev 5(1):104–111

Eisenkraft A, Falk A (2019) The use of chemical warfare agents during the Syrian Civil War. In: Chemical warfare agents: biomedical and psychological effects, medical countermeasures, and emergency response. CRC Press, Boca Raton, p 103

Ellis R (2014) Creating a secure network: the 2001 Anthrax attacks and the transformation of postal security. Sociol Rev 62(1):161–182. https://doi.org/10.1111/1467-954X.12128

Esvelt KM (2018) Inoculating science against potential pandemics and information hazards. PLoS Pathog 14(10):e1007286

Evans NG, Selgelid MJ (2015) Biosecurity and open-source biology: the promise and peril of distributed synthetic biological technologies. Sci Eng Ethics 21(4):1065–1083

FBI. Amerithrax or Anthrax Investigation. https://www.fbi.gov/history/famous-cases/amerithrax-or-anthrax-investigation

Franconi R, Illiano E, Paolini F, Massa S, Venuti A, Demurtas OC (2018) Rapid and low-cost tools derived from plants to face emerging/re-emerging infectious diseases and bioterrorism agents. In Defence against bioterrorism. Springer, Dordrecht, pp 123–139

Getz LJ, Dellaire G (2018) Angels and devils: dilemmas in dual-use biotechnology. Trends Biotechnol 36(12):1202–1205

Greer SL, Trump B (2019) Regulation and regime: the comparative politics of adaptive regulation in synthetic biology. Policy Sci 52:1–20

Gronvall G (2015) Biosecurity: the opportunities and threats of industrialization and personalization. Bull At Sci 71(6):39–44

Gronvall G (2018) The security implications of synthetic biology. Survival 60(4):165–180

Hay A (2016) The South African biological warfare program. In: Biological threats in the 21st century: the politics, people, science and historical roots. World Scientific Publishing, Singapore, pp 137–158

Kirkpatrick J, Koblentz GD, Palmer MJ, Perello E, Relman DA, Denton SW (2018) Editing biosecurity: needs and strategies for governing genome editing. George Mason University, NATO SPS ARW, July 2019

Ledford H (2010) Garage biotech: life hackers. Nat News 467(7316):650–652

Mayor A (2003) Greek fire, poison arrows & scorpion bombs: biological and chemical warfare in the ancient world. Overlook Duckworth, Woodstock

Noyce RS, Lederman S, Evans DH (2018) Construction of an infectious Horsepox virus vaccine from chemically synthesized DNA fragments. PLoS One 13(1):e0188453

Szinicz L (2005) History of chemical and biological warfare agents. Toxicology 214(3):167–181

Trump BD, Cummings CL, Kuzma J, Linkov I (2020) Synthetic biology 2020: frontiers in risk analysis and governance. Springer, Cham

Tumpey TM, Basler CF, Aguilar PV, Zeng H, Solórzano A, Swayne DE, et al (2005) Characterization of the reconstructed 1918 Spanish influenza pandemic virus. Science 310(5745):77–80

Yeh JY, Seo HJ, Park JY, Cho YS, Cho IS, Lee JH, et al (2012) Livestock agroterrorism: the deliberate introduction of a highly infectious animal pathogen. Foodborne Pathog Dis 9(10):869–877

Permissions

All chapters in this book were first published by Springer; hereby published with permission under the Creative Commons Attribution License or equivalent. Every chapter published in this book has been scrutinized by our experts. Their significance has been extensively debated. The topics covered herein carry significant information for a comprehensive understanding. They may even be implemented as practical applications or may be referred to as a beginning point for further studies.

The contributors of this book come from diverse backgrounds, making this book a truly international effort. We would like to thank all the contributing authors for lending their expertise to make the book truly unique. They have played a crucial role in the development of this book. Without their invaluable contributions this book wouldn't have been possible. They have made vital efforts to compile up to date information on the varied aspects of this subject to make this book a valuable addition to the collection of many professionals and students.

This book was conceptualized with the vision of imparting up-to-date and integrated information in this field. To ensure the same, a matchless editorial board was set up. Every individual on the board went through rigorous rounds of assessment to prove their worth. After which they invested a large part of their time researching and compiling the most relevant data for our readers.

The editorial board has been involved in producing this book since its inception. They have spent rigorous hours researching and exploring the diverse topics which have resulted in the successful publishing of this book. They have passed on their knowledge of decades through this book. To expedite this challenging task, the publisher supported the team at every step. A small team of assistant editors was also appointed to further simplify the editing procedure and attain best results for the readers.

Apart from the editorial board, the designing team has also invested a significant amount of their time in understanding the subject and creating the most relevant covers. They scrutinized every image to scout for the most suitable representation of the subject and create an appropriate cover for the book.

The publishing team has been an ardent support to the editorial, designing and production team. Their endless efforts to recruit the best for this project, has resulted in the accomplishment of this book. They are a veteran in the field of academics and their pool of knowledge is as vast as their experience in printing. Their expertise and guidance has proved useful at every step. Their uncompromising quality standards have made this book an exceptional effort. Their encouragement from time to time has been an inspiration for everyone.

The publisher and the editorial board hope that this book will prove to be a valuable piece of knowledge for students, practitioners and scholars across the globe.

Index

www.ingramcontent.com/pod-product-compliance
Lightning Source LLC
LaVergne TN
LVHW011514200926
848484LV00001B/39
* 9 7 9 8 8 8 8 3 6 0 5 7 6 *